CYBERNETICS
WITHOUT MATHEMATICS

To my teacher
Tadeusz Kotarbiński

CYBERNETICS
WITHOUT MATHEMATICS

by

HENRYK GRENIEWSKI

Translated from Polish by
Olgierd Wojtasiewicz

PERGAMON PRESS
OXFORD • LONDON • NEW YORK • PARIS

PAŃSTWOWE WYDAWNICTWO NAUKOWE
WARSZAWA

CONTENTS

FOREWORD

No one who has ever engaged in the popularization of science can regard it as an easy way of earning a living.

The basic contention is that between clarity and length. A concise exposition of the subject is usually not sufficiently clear and is, moreover, annoying to the reader since it requires considerable concentration. On the other hand, excessive length in an article, pamphlet or book may also prove exhausting to the reader. To achieve a nice balance in this respect is the initial difficulty to be overcome.

Popularization demands considerable simplification of concepts, proofs and explanations. But this involves the risk that the reader may take the author's words at their face value, and then may easily jump to the erroneous conclusion that the concepts and theorems of the discipline concerned are in fact as easy and simple as the text he is reading.

Additional difficulties arise when one has to do with a discipline as young as is cybernetics. It seems that the present state and achievements of cybernetics make it an important science, ripe for popularization. On the other hand, cybernetics

emerged only during World War II, so that in the
time scale of the history of science it is still in its
infancy. In the case of a science that is very young
and developing very rapidly — the exact condition
of cybernetics — it is somewhat difficult to ascer-
tain which of its concepts are of primary importance.

There has so far been no classic exposition of
cybernetics. In particular, its foundations have
never been systematized in any manner that has
been more or less universally accepted. This is by
no means strange, since the history of science seems
to indicate that no science has ever, in its historical
development, been built from its logical foundations. On the contrary, such logical foundations
are usually formulated *ex post*, at a late stage of
development of a particular discipline.

Yet it seems advisable to expound, or at least
to outline, an actual system of cybernetics rather
than give a mass of details and fragments somewhat obscurely inter-related.

Thus, I have decided to offer a popular but
systematic exposition of cybernetics, and even to
outline its foundations as I see them. As basic
concept, I have introduced that of a *relatively isolated system*.

I am thus flying in the face of the well-established
tradition that only such disciplines as have already
acquired certain more or less classical forms are
popularized.

I can only plead that I declared my view concerning the foundations of cybernetics at the First
International Cybernetics Congress in Namur (Bel-

gium) in 1956, and at a conference at Balaton-
vilagos (Hungary) in the same year, organized by
Professor Laszlo Kolmar on behalf of the Hungarian Academy of Sciences and devoted to mathematical logic and cybernetics. In neither case did
I meet with objections. Therefore, the present work
does not describe any universally adopted concepts,
but it does expound a view which so far has got
by without protest.

The birthplace of cybernetics was the borderland between a number of disciplines, above all
the somewhat unexpected borderland between technology, biological sciences, mathematics and mathematical logic. An intimate familiarity with that
borderland is beyond the powers of any one man.
Hence the shortcomings of my exposition.

If, in spite of the author's best intentions, the
reader finds himself grappling with a text which
is proving somewhat "indigestible", let him find
consolation in imagining the difficulties which the
author had to surmount. This piece of advice is
based on the wise maxim that nothing consoles
like other people's troubles.

A scientific book, and especially a popular one,
is as a rule something much less individual than
might seem at first glance. The author is always
a continuator of his predecessors, even if he with-
stands them; his voice is always to some extent
an echo of what he has been told by his teachers,
friends and colleagues. This book is no exception
in that respect, and its author owes a great deal
to those mentioned below.

The chapter on "Praxiological Models" would probably not have been written but for the influence of the lectures and writings of Professor Tadeusz Kotarbiński. Many ideas contained in it, also, emerged from my conversations with Professor Stefan Manczarski, Dr Jerzy Szapiro and Dr Stanisław Bogusławski.

My conversations with Professor C. Moisil (Bucharest), Dr Olgierd Wojtasiewicz and Dr Klemens Szaniawski are beyond doubt reflected in the formulations in the chapter on "Logical Models".

The idea of writing the chapter on "Economic Models" would probably not have come to me had it not been for suggestions made by Professor Oskar Lange. The concepts that chapter advances are certainly under the influence of my conversations with Professor J. Tinbergen (Rotterdam). Thanks for many valuable suggestions are due to my colleagues on the Polish Academy of Sciences Econometric Commission — the late Miss Halina Górwicz, Dr Włodzimierz Hagemajer, Dr Alojzy Chlebowczyk and Mr Władysław Sadowski.

The diagrams which — as the reader will soon appreciate — play an important rôle in this book, I owe to the ingenuity of Mrs Maria Goettig.

The translation, made by Dr O. Wojtasiewicz, was also read by Dr K. Szaniawski, and checked by Mr G. Bidwell.

My thanks are offered to all these — all contributors to this book.

_H. G._

1. BASIC CONCEPTS

1.0. Relatively isolated systems

The concept of a relatively isolated system is not new; it has been (somewhat tacitly) used in science for many centuries, at least since the time of Hippocrates' *Corpus* (cf. *References* L. 5). The need for an overt and at the same time exact use of that concept is self-evident as regards cybernetics (cf. *Ref.* G. 7). Moreover, its precise use is of importance in the logic of induction and the logic of analogies (cf. *Ref.* G. 5, G. 6, G. 9, S. 5, S. 6).

What is implied by that concept? Reference to an *"absolutely isolated system"* generally means a system which:

(1) is not influenced by the rest of the Universe, and

(2) exerts no influence on the rest of the Universe (whether any such system actually exists is not immediately relevant).

By a *"relatively isolated system"* we mean any system — and only such — which has the following two characteristics:

(1) it is influenced by the rest of the Universe, but only in certain specified ways called *inputs*, and

(2) it influences the rest of the Universe, but only in certain specified ways called *outputs*.

These conceptions, originally so simple, have to be subjected to certain complications: the influence of the system upon itself (e. g., self-correction) has to be taken into consideration, involving acceptance of the fact that some (but not all) outputs of the system may at the same time be inputs (feedback coupling).

The first basic concept considered, then, is that of a *relatively isolated system*. The concept of such a system, the notions of *input* and of *output* are abstract notions somewhat difficult to formulate with adequate precision. They can be formulated precisely if they are treated as what are called "primitive terms", and if a certain set of postulates which infuse meaning into those concepts is adopted. That is to say, they can be formulated precisely by the method used in elementary geometry to give precision of meaning, e. g., to the concepts of point and plane. For our purposes, however, such postulates would be over-complicated; better to do with a somewhat simplified explanation, closing our eyes to certain essential difficulties.

1.1. Calendar, repertory, trajectory

Every *input* and every *output* of a given relatively isolated system is associated with:

(1) its *calendar*, i. e., a certain set of moments, or intervals of time, of at least two elements, and

(2) its *repertory*, i. e., a certain set of distinguishable states.

In a given system, every *input* and every *output* adopts one, and only one, distinguishable state in any moment of its calendar.

The function establishing a relation between the elements of the calendar of a given input (output), and the distinguishable states belonging to the repertory of that input (output) is called the *trajectory* of that input (output). By a "trajectory" is usually meant the path of a missile, i. e., a function which establishes a relation between moments of time and positions of the missile in space. In the more generalized meaning used in this book, a trajectory is a function establishing a relation between the various elements of time (moments or intervals) and the various distinguishable states of the given input (output) in a certain abstract space, namely the repertory of distinguishable states.

1.2. Stimulus and reaction

Instead of using the expression "a distinguishable state of an *input*", we may use briefly "stimulus", and "a distinguishable state of an *output*" we may shorten to "reaction".

The term "trajectory", defined above, is borrowed from mechanics, but its meaning has, as indicated, been greatly generalized. In a similar way, the terms "stimulus" and "reaction" are borrowed from physiology or psychology, but in meaning they have been considerably generalized. Such generalization of meanings of terms borrowed from languages used by specialized disciplines seems to be a useful method of establishing cybernetical terminology.

1.3. Four types of systems

Four types of relatively isolated systems will be introduced: differentiation will be made between *reliable* and *unreliable* systems, on the one hand, and *prospective* and *retrospective* systems, on the other, each of these dichotomic classifications being independent of the other.

It is reasonable to ask: (1) What is the difference between inputs and outputs? (2) Is there any relation between stimuli and reactions? Answers to these questions will be given separately for systems of each type.

1.4. Prospective reliable systems

These systems have the following two characteristics:

(1) The repertory of each input consists of at least two distinguishable states.

(2) The present distinguishable state of any *output* is always *univocally* determined by past and present distinguishable states of all the inputs of the given system.

Condition (2) is called the principle of local determinism.

Now for two examples of prospective reliable systems.

(1) A system consisting of a key and a lock is probably one of the simplest examples: the key is obviously the only input, and the bolt of the lock, the only output. The state of the bolt is always univocally determined by the past state (movement) of the key.

(2) Part of a well-functioning electrical installation, connected with the source of electric power and consisting of a switch, two conducting wires, and an effective electric bulb properly attached to the wires. The switch is the only input, and the bulb, the only output. The state of the bulb is always univocally determined by the (present) state of the switch.

1.5. Prospective unreliable systems

Such systems have the following two characteristics:

(1) The repertory of each *input* consists of at least two distinguishable states.

(2) The present distinguishable state of any input is always determined by past and present distinguishable states of all the *inputs* of the given system, with a constant probability greater than 50 per cent.

Condition (2) is called the principle of local pseudodeterminism.

The following example is none the less instructive for being very simple: There are two urns on the table (label them briefly the "right" urn and the "left"). Each urn contains only white and black balls; in the left urn are a majority of black balls, and in the right urn a majority of white ones. There is also a plate on the table. Balls are repeatedly drawn from the urns by plunging a hand — arbitrarily — sometimes in the right and sometimes in the left. Only one ball at a time is drawn out of a single urn. The ball drawn is always put on the plate, but replaced, before the next draw takes place, in the urn from which it was drawn. This is a prospective unreliable system. The input is the hand which draws the balls. The output is the state of things on the plate. The repertory of the input consists of two distinguishable states: movement of the hand reaching to the left urn and movement of the hand reaching to the right urn. There are also two distinguishable states of the output: the appearance on the plate of the black ball and of the white. If the hand reaches to the left urn, that involves a probability greater than 50 per cent that a black ball will appear on the plate. If the hand reaches to the right urn, that involves a probability greater than 50 per cent that a white ball will appear on the plate.

Philosophical determinism seems to imply that every prospective system is *reliable*, whereas our everyday experience seems to indicate that every prospective system is unreliable. This apparent contradiction can be easily explained in terms of philosophical determinism: in practice we seldom know all the inputs of a given prospective reliable system; because it is thus distorted by our ignorance, that system seems to us to be unreliable. There is a widely known example, due to Bertrand Russell (cf. *Ref.* R. 2), to show the fallibility of the concept of a determined sequence of events: a coin is put into a ticket machine, but a sudden earthquake prevents the machine from delivering the ticket. By reference to that example, a prospective reliable system might facetiously, though not very precisely, be likened to a slot machine which has received a guarantee that no earthquake will occur. And even if the guarantee is not one hundred per cent dependable, yet reliable to a high degree, the system under consideration is a prospective one and unreliable but can easily be approximated by a prospective reliable system. On the other hand, Russell's slot machine without a guarantee is a prospective unreliable system.

1.6. Reaction time

We usually have to do with prospective systems, whether reliable or otherwise, in which the effect of past and present states of inputs on the present states of the outputs, as described in 1.4 and 1.5, is limited as follows:

To every paired input and output of a specified system a certain non-negative number of time units is ascribed, representing the time necessary for the reaction to take place — called reaction time or *time-lag*. It is also assumed that the actual distinguishable state of any output is not influenced by any past more remote than the *time-lag* ascribed to the paired input and output.

1.7. Retrospective reliable systems

These systems have the following two characteristics:

(1) The repertory of every *output* consists of at least two distinguishable states.

(2) Any past (but sufficiently remote from the present) distinguishable state of any *input* is always *univocally* determined by present and past (but not prior to the input state in question) distinguishable states of all *outputs*.

Condition (2) is called the principle of local paradeterminism.

It might be maintained that Sherlock Holmes's talent consisted in his ability to discover retrospective reliable systems within the field of his investigations. Knowing the "today's" states of outputs (i. e., the traces of a crime), he univocally determined the "yesterday's" states of inputs (i. e., the person of the criminal and his methods of action).

1.8. Retrospective unreliable systems

These systems have the following two characteristics:

(1) The repertory of every *output* consists of at least two distinguishable states.

(2) Any past (but sufficiently remote from the present) distinguishable state of any *input* is always determined with a probability greater than 50 per cent by present and past (but not prior to the input state in question) distinguishable states of all the outputs.

Condition (2) is called the principle of local para-pseudodeterminism.

Still using terms of crime fiction it might be said that a retrospective reliable system is every detective's dream; reality, however, falls short of dreams, so that all detectives in their investigations have to do almost entirely with retrospective unreliable systems.

1.9. Determinators and paradeterminators

By a *"determinator"* of any *output* of a given prospective system is meant a function which assigns reactions to stimuli.

By a *"paradeterminator"* of any *input* of a given retrospective system is meant a function which assigns stimuli to reactions.

Both determinators and paradeterminators can be represented either in an analytic way or by means of matrices. Such matrices will be discussed later, in connection with binary (zero-one) systems.

1.10. Duality

If in any system *the direction of time is reversed,* the inputs replaced by outputs and the outputs by inputs, then a *prospective* system becomes a *re-*

trospective one (and *vice versa*, a retrospective system becomes a prospective one). In prospective systems, the present state of the outputs is determined (logically or probabilistically) by the past and present states of the inputs; in retrospective systems, the situation is the opposite: the past state of the inputs is determined by the recent past and present states of the outputs.

The theory of relatively isolated systems is marked by duality: every theorem referring to the prospective systems has one, and only one, corresponding (dual) theorem referring to the retrospective systems, and *vice versa*; if one of such two theorems is proved, the other can be proved by simple transformations.

Duality is characteristic of many deductive theories, such as the algebra of distributive lattices, Boolean algebra and projective geometry.

1.11. Relative isolation hypothesis

On one occasion, Professor Kotarbiński asked a very awkward question — How do we know that a given object is a relatively isolated system? There is, of course, no difficulty in deciding that a certain object which no longer exists and which was adequately studied while it existed, was a relatively isolated system. Professor Kotarbiński ·was not, however, thinking of such a trivial case but rather of the case in which the object in question is to exist in the future, or the case in which it escaped our observation in the past.

To answer this question fully, one would have to be familiar with the detailed history of the natural and the social sciences. The present author's lack of qualification in that respect restricts him to the comment:

(1) Mankind has for thousands of years been resorting to certain concepts of a prospective system, because since time immemorial people have wanted to know how or by what means they could obtain certain desirable results. In the course of time, some inputs were struck out of certain systems (e. g., people stopped believing that apparent configurations of heavenly bodies influenced the fate of men), and some new inputs were added to other systems (e. g., the pathogenic rôle of bacteria).

(2) Mankind has for thousands of years, also, resorted to certain concepts of a retrospective system in the desire to know, what unknown factor or factors in the past has or have brought about the present state of things (e. g., who it was who committed a murder, robbery, etc.). The lapse of time, new experience, a shrinkage or expansion of the sphere of superstition, the progress of science — all these have contributed to the striking out of some outputs of retrospective systems and to the adding of new ones.

The conviction that a certain object is a relatively isolated system seems usually to be born of the collapse of the conviction that a certain similar object was such a system.

It might also be said, on a much lower plane: we do not know how the conviction is born that a given object is a prospective, or retrospective, system; let us, however, assume that it is a relatively isolated system, since this fertile hypothesis enables us to prove many theorems, helps us in empirical and experimental research, and has one fundamental advantage — although it cannot be fully proved, it can be experimentally disproved.

(Cf. *Ref.* G. 5, G. 6, G. 9)

2. BASIC CONCEPTS (CONCLUDED)

2.0. Informed, informing and information systems

Among all the possible kinds of inputs (outputs) we distinguish the two extremes:

(1) information inputs (outputs),
(2) physical inputs (outputs).

By *"information"* we imply any message, any communication, any permission, any order and any prohibition. We say that a given input (output) is an *information* input (output) when each of its distinguishable states is an information. (The reader must be warned that the term "information" is used, both colloquially and in cybernetical terminology, also in other meanings.) We say that a given input (output) is a *physical* input (output) when none of its distinguishable states is an information.

In some cases, it may be very difficult to distinguish between an information input (output) and a physical one.

The concept of information input (output) makes it possible to introduce three important concepts:

(1) informed system,
(2) informing system,
(3) information system.

An informed system is a relatively isolated system having at least one information input.

An informing system is a relatively isolated system having at least one information output.

An information system is both an informed and an informing system.

Receptors of a nervous system can be quoted as classical examples of information inputs; effectors of a nervous system are equally classical examples of information outputs; and any nervous system as a whole serves as an example of an information system.

Modern technology has given rise to many types of information systems, now indispensable in a civilized society. They might be classified as information systems used to transmit information (telephone, teleprinter), information systems used to record information (dictaphone), and information systems used to process information (logical, mathematical, statistical and book-keeping machines).

2.1. The graphical method

With a view to simplifying all the explanations as much as possible, we shall henceforth confine our analysis to prospective reliable systems. Whenever the term "system" is used without any other qualifications, a prospective reliable system is meant.

A graphical method of representing systems under discussion will be introduced: any system in question, in which no partial systems (to be dis-

cussed later) are distinguished, will be shown as
a rectangle with sides erected vertically and hori-
zontally. Each *input* will be marked as a straight
or polygonal line which has one point on the *upper
horizontal side* of the rectangle, and which bears
an arrow directed towards the rectangle. Each
output will be marked as a straight or polygonal

FIGURE 2.1.0

Systems having different number of inputs and outputs

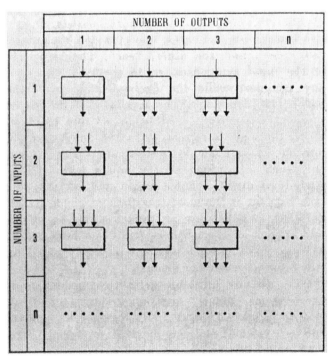

line having one point on the *lower horizontal side*
of the rectangle, and bearing an arrow directed
away from the rectangle (see Fig. 2.1.0).

2.2. "Zero-one" systems

The repertory of any input (output) can consist
of a small, a vast, or sometimes even an infinite
number of distinguishable states. The simplest
situation is when such a repertory consists of only
two distinguishable states, one of which we usually
call *"zero"*, and the other, *"one"*. The zero-state
of the *input* symbolizes, so to speak, the lack of
any stimulus, while the one-state represents the
only stimulus possible. The zero-state of the *output*
symbolizes, so to speak, the lack of any reaction,
while the one-state represents the only reaction
possible. A zero-one system is any system — and
only such — in which all the inputs take on precisely two distinguishable states and all the outputs take on at the most two distinguishable states.
Zero-one systems play an immense rôle in cybernetics; some of them will be studied in brief below.
The theory of zero-one systems is largely based on
two-valued sentential calculus.

We must now introduce certain elementary zero-
one systems, namely negation systems and delay
systems (Figs. 2.2.0 and 2.2.1) together with alternative systems and conjunction systems (Figs.
2.2.2 and 2.2.3).

A *negation system* is a system of one input and one output in which reaction appears at the output if, and only if, there is no stimulus at the input, and in which there is no reaction at the output if, and only if, there is a stimulus at the input. A relay structure of such a negation system involves no difficulties: the input is an electric circuit with a switch, the circuit including an electromagnetic coil; the output is another electric circuit with a switch controlled by the electromagnet. The relay (i. e., the electromagnet plus the controlled switch) is so built that when the input-circuit is open, the output-circuit is closed, and vice versa.

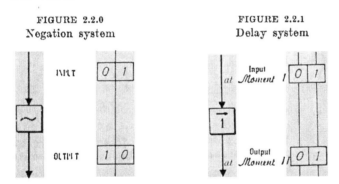

FIGURE 2.2.0
Negation system

FIGURE 2.2.1
Delay system

A *delay system* resembles a bureaucrat who does nothing except hold up all decisions. The output state here is the same as was the input state one time unit earlier (Fig. 2.2.1). Delay systems are priceless elements in the construction of "memory" devices for storing information.

An *alternative system* is an "hypersensitive" system such that it is sufficient for a stimulus to act on one of the inputs to produce a reaction at the only output (Fig. 2.2.2). In a relay model of such a system, we have to do with three electric circuits, two of which are the inputs, and the third, the output. Each input-circuit has a switch and includes a coil of a separate electromagnet. These electromagnets control the two switches of the output-circuit which are connected parallely.

FIGURE 2.2.2   FIGURE 2.2.3
Alternative system   Conjunction system

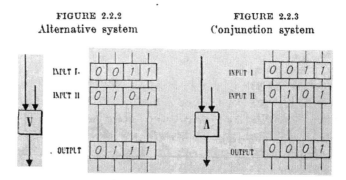

A *conjunction system* might be called hyposensitive: reaction appears at the only output if, and only if, stimuli act on both inputs (Fig. 2.2.3). In a relay model of such a system, the structure resembles that of a model of an alternative system, the only difference being that the two switches of the output-circuit are connected serially, and not parallely.

(Cf. *Ref*. B. 1, C. 1, G. 3, G. 5, GBM, M. 2, M. 3, S. 4)

2.3. Replicating systems

We say that a given system replicates n times rather than that it satisfies all the following conditions:

(1) it has only one input,

(2) it has n outputs,

(3) all the outputs have the same trajectory,

(4) the only output trajectory is identical with the trajectory of the only input or differs from the latter only by being shifted in time.

FIGURE 2.3.1
Replicating systems

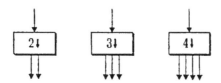

The common mimeograph is an elementary example of a replicating system.

The zero-one matrix of a system which replicates twice might be shown as in Table 2.3.0.

TABLE 2.3.0

State of the input at Moment I	State of the first output	State of the second output
	at Moment II	
0	0	0
1	1	1

A special symbolism is introduced for replicating systems (Fig. 2.3.1).

2.4. Algebra of systems

The theory of systems bears a considerable resemblance to an algebra. By an "algebra", in the modern interpretation of the term, is meant such, and only such, a theory as is built in the following way:

There is a certain given non-empty set of elements at least some of which are effectively given; we study certain operations performed on all or certain elements of the set in question; as a rule, every such operation gives a result which again belongs to the set in question.

In the theory of systems, the given set is of the family of all systems; some of them (never mind, for the moment, which) are effectively given. In this theory, we deal with operations performed on relatively isolated systems; in almost all cases the result of such an operation or operations is a new relatively isolated system.

Of all the operations of the algebra of relatively isolated systems we shall deal only with binarization and coupling.

(Cf. *Ref.* G. 2, G. 3, G. 11, M. 2, P. 1, S. 4)

2.5. Binarization of inputs and outputs

We shall briefly discuss here a certain numerical relationship between the number of distinguishable states of a given input (output) and the number of inputs (outputs) of a given system. That question, of great importance in information theory, will be illustrated here by means of a few examples only.

Let us consider a system of only one input, the repertory of which consists of four distinguishable states denoted by the following digits:

$$0, 1, 2, 3.$$

It can easily be observed that instead of studying such a system we may study a certain substitutive system of *two* inputs, each of which has a repertory consisting of only two distinguishable states: *0* and *1* (zero-one inputs). The following Table explains why this is so (Table 2.5.0):

TABLE 2.5.0

Before binarization	After binarization	
State of only input	State of	
	input I	input II
0	*0*	*0*
1	*0*	*1*
2	*1*	*0*
3	*1*	*1*

Let us again consider a system of only one input, the repertory of which consists of eight distinguishable states marked by the following digits:

$$0, 1, 2, 3, 4, 5, 6, 7.$$

Here, too, it can easily be observed that instead of studying such a system we may study a certain substitutive system of three zero-one inputs (Table 2.5.1).

TABLE 2.5.1

Before binarization	After binarization		
State of only input	State of		-
	input I	input II	input III
0	0	0	0
1	0	0	1
2	0	1	0
3	0	1	1
4	1	0	0
5	1	0	1
6	1	1	0
7	1	1	1

Generally: instead of considering an input (output) with a repertory of 2^n distinguishable states we may consider n inputs (outputs), each with a repertory of two elements — zero and one — or, in other words, a binary repertory.

Let us imagine an information input through which we convey, letter by letter, some information in a language the alphabet of which (i. e., letters plus punctuation marks) consists of 2^n elements. In view of what has been said above, that single information input can be replaced by n zero-one information inputs through which the same amount of information can be conveyed without any difficulty.

All this indicates the great usefulness of zero-one (binary) systems in cybernetics, since there is a large family of systems which can be reduced to zero-one systems.

Their importance is the greater since:

(1) the building of zero-one systems (using various techniques, e. g., relays, vacuum tubes, ferrites, etc.) presents no serious difficulties;

(2) the functioning of nervous systems — in the opinion of those physiologists who stand for the "all-or-nothing" principle — is usually binary in character (i. e., in a given neurone either there is an impulse, which is interpreted as "one", or there is no impulse, which is interpreted as "zero").

(Cf. *Ref.* B. 2, M. 2, R. 1, S. 1, S. 2)

2.6. Serial coupling

Let us consider two systems, *I* and *II* (see Fig. 2.7.0, upper diagram). We suppose that one of the *outputs* of system *I* is at the same time an *input* of system *II*, and further that the trajectory of such output is identical with the trajectory of such input. Under such conditions, systems *I* and *II* form a whole, let us call it "*U*", which again is a system. We say in such a case that the system *U* is a direct *serial coupling* of systems *I* and *II*. Graphical explanation of this definition is shown at Fig. 2.7.0, upper diagram.

And here is an example of a serial coupling: a typist is copying something. This fragment of reality, apparently so simple, includes as many as three systems:

System I = the typist. Inputs = the typist's eyes (other receptors can be disregarded here). Outputs = the typist's fingers (other effectors can be disregarded here). Each state of each output (i. e., each purposive movement of each finger) is univocally determined by the state of the inputs (under the assumption that the typist makes no mistakes).

System II is somewhat strange: the inputs here are the typist's fingers, the outputs are the keys of the typewriter. Each state of each ouput is univocally determined by the state of the inputs.

System III = the typewriter. Inputs = the keys. Outputs = the type letters. Each state of each output is univocally determined by the state of the inputs.

Now notice that:

(1) each output of system I is an input of system II, the output trajectory being always identical with the input trajectory,

(2) each output of system II is an input of system III, the output trajectory being always identical with the input trajectory.

Hence:

(1) system I is directly coupled serially with system II, and

(2) system II is directly coupled serially with system III.

In addition to direct serial couplings, there are also indirect serial couplings. We say that system I is indirectly coupled serially with system III rather than that there exists a system II satisfying the following two conditions:

(a) system I is directly coupled serially with system II,

(b) system II is directly coupled serially with system III.

An example of an indirect serial coupling is that of our typist who is indirectly coupled serially with her typewriter.

2.7. Feedback coupling

It sometimes happens that not only is system I directly coupled serially with system II, but also system II is directly coupled serially with system I (see Fig. 2.7.0, middle diagram); in such a case we have to do with a direct *feedback coupling*. In other words, it sometimes happens that a system U satisfies both the following conditions:

(1) U is a serial coupling of systems I and II,

(2) U is a serial coupling of systems II and I.

System U is called a direct feedback coupling of systems I and II if, and only if, the above conditions are satisfied.

In addition to direct feedback coupling, there are also indirect feedback couplings. We say that system I is indirectly feedback-coupled with system II rather than that only one of the following three conditions is satisfied:

either (1) system I is indirectly coupled serially with system II, and system II is directly coupled serially with system I;

or (2) system *I* is directly coupled serially with
system *II*, and system *II* is indirectly coupled
serially with system *I*;

FIGURE 2.7.0

Serial, feedback and parallel couplings

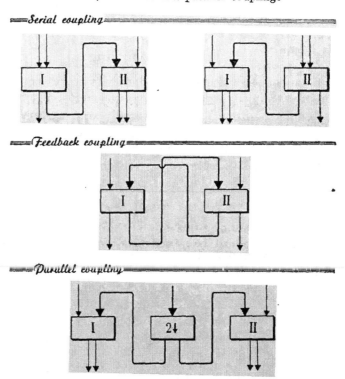

or (3) system *I* is indirectly coupled serially with
system *II*, and vice versa, system *II* is indirectly
coupled serially with system *I*.

This somewhat complicated definition becomes easier to grasp by reference to Fig. 2.7.1.

FIGURE 2.7.1

Indirect feedback couplings

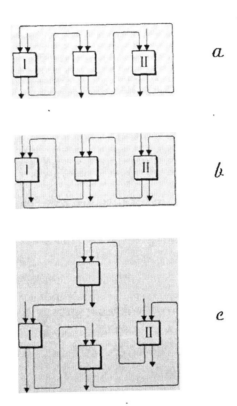

Feedback coupling plays an immense rôle in inanimate and animate nature, in technology and in social organization.

A good example of a feedback coupling is that of a pair of systems consisting of a planning unit and its executive unit. The former is coupled serially with the latter (transmission of orders resulting from the plan), and the latter is coupled serially with the former (transmission of reports on the execution of the orders received).

Another example, taken from the field of social relations, is that of the couple: teacher-pupil. The teacher is coupled serially with his pupil (conveying of knowledge, asking questions), but the pupil, too, is coupled serially with his teacher (replying to questions, asking questions). The teacher's behaviour influences that of his pupil, but the pupil's behaviour also influences that of his teacher.

Hence:

(1) the teacher influences his own behaviour in an indirect way, viz., through his pupil,

(2) the pupil also influences his own behaviour in an indirect way, viz., through his teacher.

Other examples of feedback coupling will be discussed later.

Two kinds of feedback couplings will be distinguished:

(1) negative feedback,

(2) positive feedback.

(This is not an exhaustive classification; there are feedback couplings which are neither negative nor positive).

(Cf. *Ref.* S. 3, L. 3, W. 2)

2.8. Negative feedback

This time we begin with an example. A man is driving a car on a clear road, looking at the speedometer from time to time. He wants to drive at a more or less uniform speed, so, according to the speed shown by the speedometer, he either increases or reduces his pressure on the accelerator. The driver is coupled serially (but indirectly) with the car (right foot — the accelerator), and the car is coupled serially (but also indirectly) with the driver (the speedometer — the driver's eyes). When the speed of the car exceeds what the driver considers standard, he reduces the pressure on the accelerator; when it falls below that standard (say, 50 m. p. h.), he increases the pressure on the accelerator.

Let us now generalize the above example: There are two feedback-coupled systems, I and II. That output of system I (called "W_I") by which that system is coupled with system II has a repertory of distinguishable states which are quantitative in nature. One of these distinguishable states (not an extreme one, but otherwise any of such states) will be called "equilibrium state". That output of system II (called "W_{II}") by which system II is coupled with system I has a repertory of distinguishable states also quantitative in nature. If the actual distinguishable state of output W_I differs from the equilibrium state, then output W_{II} takes on such a distinguishable state as, acting through system I, brings the next state of W_I closer to the equilibrium state. If all the above conditions are satisfied, we say that system *I* is feedback-coupled negatively with system *II*.

FIGURE 2.8.0

Negative feedback coupling

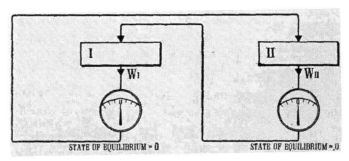

STATE OF EQUILIBRIUM = 0   STATE OF EQUILIBRIUM = 0

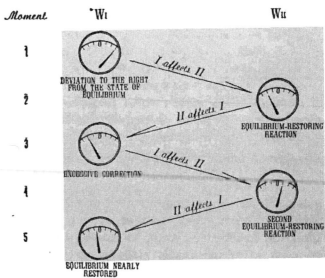

This terminological explanation can be generalized
still more: it is not necessary to assume that the
distinguishable states of outputs W_I and W_{II} are
quantitative in nature. We can assume much less,
namely that both the repertory of W_I and the
repertory of W_{II} are metric spaces, but maybe to
introduce here the concept of metric space would
mean getting too involved in mathematical techni-
calities.

Negative feedback is a means of maintaining
a state not remote from that of equilibrium, a means
often resorted to in nature, in technology and in
organizational matters.

2.9. Positive feedback

To begin once more with an example. Suppose
that two persons are talking, and that neither of
them is particularly phlegmatic. The subject of
their conversation is somewhat touchy for both.
The conversation begins in a rather low tone. The
opening sentence is spoken by person I quite calmly,
but it irritates person II who replies a little more
loudly; person I in turn replies still more loudly,
and the conversation ends in each of them trying
to shout the other down.

This, too, can be generalized: There are two
systems, I and II; they are feedback-coupled. That
output of system I (called "W_I") which couples I
with II has a repertory of distinguishable states
which are quantitative in nature. One of these

states (not the maximum one, but otherwise any of such states) will be called "equilibrium state". That output of system II (called "W_{II}") which

FIGURE 2.9.0

Positive feedback coupling

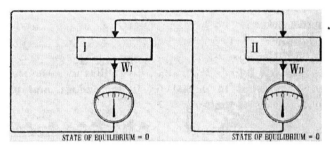

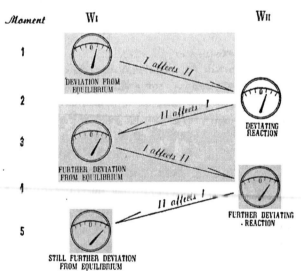

couples II with I also has a repertory of distinguishable states which are quantitative in nature.

If the actual distinguishable state of W_I differs from the equilibrium state, then W_{II} takes on such a distinguishable state as, acting through I, further removes (if that is still possible!) the subsequent state of W_I from the equilibrium state. If all these conditions are satisfied, we say that system I is feedback-coupled positively with system II (see Fig. 2.9.0).

This terminological explanation is not quite general since it is not applicable when we have to do with a continuous calendar (i. e., one in which no moment of time has a direct predecessor or a direct successor). Positive feedback could be defined in more general terms by using the concept of metric space, but we have decided to eschew such technical details.

Many biological and social phenomena are based on positive feedback.

2.10. Parallel coupling

The concept of parallel coupling can be explained by means of the concepts of serial coupling and replicating system.

Let us consider two systems, I and II (see Fig. 2.7.0, lower diagram). We say that system U (the whole lower diagram of 2.7.0) is a *parallel coupling* of systems I and II rather than that there is a *replicating* system U^* which satisfies both the following conditions:

(1) U^* is coupled serially with I,

(2) U^* is coupled serially with II.

The concept of parallel coupling has many applications in technology. A simple example of an (information) parallel coupling is that of two persons reading the same newspaper (cf. *Ref.* G. 7 and GBM).

2.11. Self-coupling

It happens that a given system is coupled serially with itself, i. e., some of its outputs are at the same time among its inputs. In such — and only such — cases, we say that the system in question

FIGURE 2.11.0

Self-couplings

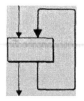 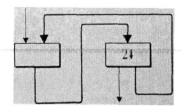

is self-coupled. Self-coupling is, of course, a special case not only of serial coupling, but also of feedback coupling (see Fig. 2.11.0).

2.12. Inputs, outputs, throughputs

If two systems, *I* and *II*, are coupled serially in such a way that *only one* output of *I* is an input of *II*, we call that sole *output* of *I* the throughput from *I* to *II* (cf. *Ref.* G. 11).

2.13. Matrices of couplings

Now consider any non-empty and finite set of systems *Z*. We assume that any two systems belonging to *Z* have the characteristic that if system *I* is directly coupled serially with system *II* they are coupled by a throughput, i. e., by a single output-input. We further assume that any system belonging to *Z* has, apart from throughputs, at most one input which is not a throughput, and at most one output which is not a throughput (they are the so-called inputs from the outside and outward outputs of *Z*).

In such a case, the whole set can be described by means of a zero-one matrix (table) built thus:

(1) one and only one line of the matrix is allotted to each system belonging to *Z*,

(2) one and only one column of the matrix is allotted to each system belonging to *Z*,

(3) the square where the line corresponding to system *A* intersects the column corresponding to system *B* has "*0*" if *A* is not coupled serially with *B*, and "*1*" if there is such a serial coupling between them.

A set Z consisting of three systems, A, B and C, will be analysed by way of example (see Fig. 2.13.0).

FIGURE 2.13.0

The set Z of systems A, B and C

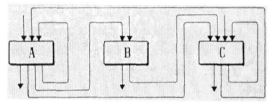

The matrix of couplings in this case will be as follows (Table 2.13.1):

TABLE 2.13.1

Input	Output				Total of outputs
	from the outside	A	B	C	
to the outside	·	1	1	1	3
A	1	1	0	1	3
B	1	1	0	0	2
C	1	1	1	1	4
Total of inputs	3	4	2	3	12

It can easily be observed that in a matrix of couplings a couple of "*1*", situated symmetrically with respect to the top-left-bottom-right diagonal of the matrix, corresponds to a direct feedback coupling (unless an outward output or an input from the outside is in question), and that "*1*" appears on that diagonal if, and only if, a given system is self-coupled.

3. SUBJECT MATTER AND METHOD

3.0. Science and metascience

If we study a section of reality we describe it by means of certain sentences which belong to a certain language, called "*object language*". If the set of sentences is sufficiently rich and logically coherent, and embraces sentences which have fairly general validity, we may be entitled to consider that set of sentences a science. Let us assume for the time being that *only* such a set of sentences is called a "*science*".

Let us now disregard that sector of reality which is described by the science and concentrate on the science itself. A number of questions can be raised here. Is that science internally consistent? Is it experimentally verifiable? Which sentences of that science follow logically from a given sentence? Which sentences of that science are logically independent of one another? Contrary to superficial appearances all these questions, and all answers to them, are formulated not in the object language, but in another language (which includes, among other things, the *names* of the sentences belonging to the science). That language, serving the new purpose of describing the *science*, will be called "*metalanguage*". And the answers to the questions

4

asked a moment ago will belong to a field different from *science* as we narrowly interpret it. They will belong to a certain set of sentences which will be called *"metascience"*. In this way, modern methodology differentiates between logic and metalogic, arithmetic and meta-arithmetic, geometry and meta-geometry, etc. (cf. *Ref.* G. 4 p. 134, and M. 3).

One does not usually engage in metascience before engaging in science — but with one notable exception with which we shall deal later on.

The answer to the question, what is the subject matter of a given science, is *not* a matter for that science, but for metascience. When, before becoming familiar with at least one sentence belonging to a given science, we ask about the subject matter of that science, our question, although not improper, is a premature one: we engage too early in metascience. That is why it seems reasonable to engage in a given science first and subsequently in metascience. This is the order observed in this book, and for this reason the question of the subject matter of cybernetics will be tackled only at this stage. The further step will also be on the metascientific level: we shall deal with the methods of cybernetic research.

3.1. Matter of which systems are built

In this section, the term *"technical matter"* implies *"matter purposively shaped by man or by a group of men"*, whereas *"non-technical matter"* implies "matter which is not purposively shaped by any human being".

There are relatively isolated systems which are built of inanimate non-technical matter, i. e., systems shaped by nature out of inanimate matter alone. Further, there are systems which are built of animate matter: any organism, any organ, any nervous system can be treated as a relatively isolated system. (In the case of the nervous system we treat receptors as inputs and effectors as outputs.) The pupil is but a relatively isolated system, with receptors as inputs and effectors as outputs. The teacher, by giving instruction and asking questions, induces certain states of certain inputs (receptors) of the system in question, his purpose being that of bringing about certain specified states of outputs (effectors) in the pupil, i. e., obtaining correct answers. This object is not always achieved, since it is well known that the teacher (a) has only a limited possibility of controlling the inputs of his pupil, and (b) even if he has an exemplary knowledge of psychology, he does not know exactly the law which describes the state of the outputs as a certain function of the states of the inputs (insufficient knowledge of the determinators involved. See Section 1.9). Of course, the pupil is not the only example of a relatively isolated system. The same holds true of the patient, and the difficulties which the physician has to face are strikingly analogous to those of the teacher.

But let us go further. There is an immense variety of systems consisting of technical matter: all sorts of machines, measuring instruments, and, as a rule, the experimental objects of physics. Any physicist will see immediately that practically every experimental system used for teaching purposes is a relatively isolated system.

Still further: there are other systems, still more complicated and qualitatively different from those already mentioned — social systems; one of the simplest examples here is that of a perfectly disci-plined detachment of an army during manoeuvres; much more complicated examples are offered by political economy.

In our further analysis we shall distinguish five kinds (as it were development stages) of matter, or, more strictly speaking, five kinds of fragments of matter:

(1) A — inanimate non-technical matter,
(2) B — animate non-technical matter (individual plants and animals, but not human beings),
(3) C — individual human beings,
(4) D — inanimate technical matter,
(5) E — animate technical matter.

A above will include, e. g., any rock not shaped by man, but will not include any man-shaped rock, or any tool. B will include, e. g., any dog or pigeon not subjected to any surgical operation by man, but will not include, e. g., any dog which has had its cerebral hemispheres removed surgically; B will further include no complex of plants or animals, although it does include any individual plant and individual animal. C will not include any society, any social class, any staff, or any family, although it will include any individual man belonging to any society, social class, staff or family. *D* will include any building, any ship, any machine and any weapon, but it will not include any stone not shaped by man, even if such stone were used by a man to kill another man or an animal. *E* will include live tissue separated from a living organism in a laboratory and artificially kept alive, as well as, e. g., any animal with its cerebral hemispheres surgically removed.

Qualitative differentiation between the various kinds, or development stages, of matter is known both in every-day life and in philosophy, especially Marxist philosophy. But a division which differentiates between inanimate non-technical matter and inanimate technical matter, and between animate non-technical matter and animate technical matter is less common. It seems, however, that such a division, more detailed than those commonly used, is valuable to all those who want to engage in metacybernetics or the philosophy of cybernetics. Although relatively isolated systems can be found in inanimate non-technical matter (e. g., a piece of rock-crystal which refracts the sun's rays), and although the literature of the subject has long since been giving examples of positive feedback in inanimate nature (cf. *Ref.* L. 3), yet inanimate technical matter is much more interesting from the cybernetical point of view than is inanimate non-technical matter. Experience shows that men can by their conscious activity impart to inanimate technical matter certain characteristics

that are alien to inanimate non-technical matter,
though not alien to animate matter, individual
human beings and human groups. In subsequent
chapters, devoted to the construction of models,
these questions will be discussed in greater detail.

It must be explicitly stated that the division of
fragments of matter as performed above into the
kinds A, B, C, D and E is *not* a classification, since
classification must be exhaustive and disjunctive.
Never mind now whether our division is disjunctive
or not, suffice it to say that it is not exhaustive.

Let us notice now that within each kind — A,
B, C, D and E — we can find many relatively
isolated systems. The reader can easily do so for
himself (an example of a system belonging to A
has been given above). But we can with equal ease
find relatively isolated systems which cannot be
included in any of the five kinds enumerated. If
we consider a relatively isolated system formed as
a result of an indirect serial coupling between
a man and a machine (the typist and the typewriter) we see that part of it belongs to C and the
rest to D. Thus it appears that our division is not
exhaustive since there exist fragments of matter
which are relatively isolated systems and which
do not fit into any of the five kinds as specified
above.

In order to approach more nearly to the ideal
of the exhaustive division, let us introduce new,
combined, kinds of fragments of matter. If a relatively isolated system belongs to kind X and
another such system to kind Y, the serial coupling

of the two systems will be included in the new kind,
namely (X, Y), which will be considered identical
with (Y, X). The list of these new kinds is given

TABLE 3.1.0

Kind of matter	A	B	C	D	E
A	A	.	.	.	.
B	(A,B)	(B,B)	.	.	.
C	(A,C)	(B,C)	(C,C)	.	.
D	(A,D)	(B,D)	(C,D)	D	.
E	(A,E)	(B,E)	(C,E)	(D,E)	.

in Table 3.1.0. We have thus 12 new kinds of fragments of matter, among them, for example:

(1) (A, B) — e. g., feedback coupling of an animal
and its inanimate non-technical surrounding,

(2) (B, B) — e. g., feedback coupling of two animals,

(3) (A, C) — e. g., feedback coupling of man and
his inanimate non-technical surrounding,

(4) (B, C) — e. g., serial coupling of a bareback
rider, and his horse,

(5) (C, C) — e. g., serial or feedback coupling of
two men,

(6) (A, D) — e. g., serial coupling of the drill and
the rock, etc.

It is obvious that the division of fragments of
matter into

$$5 + 12 = 17$$

is still *not* exhaustive. Further kinds, such as for
instance (X, Y, Z), can be formed. One of such
new kinds, namely (B, C, D), would include any

system which is a serial coupling of the rider who sits a saddle, and his horse.

This question will not be discussed here in detail. The purpose of these remarks has been to draw the reader's attention to the risks attendant on all generalizations concerning relatively isolated systems and the kinds of matter, and to the care needed in studying that problem.

3.2. The subject matter of cybernetics

It seems reasonable to risk the following statement: The subject matter of cybernetics involves only relatively isolated systems, in particular the informed, informing and information systems (see Section 2.0).

Does this conform to the common view that cybernetics is a general science of control and communication? More or less — yes. It should first be explained, however, that control and communication are practically identical if given exact meanings.

What is meant by "communication"? Conveying information, of course. And what is meant by "control"? Control also consists in conveying information intended to produce desired changes. Thus, all control is communication. But, on the other hand, all communication is control, since all communication consists in conveying information intended to result at least in such changes as arise from the fact that information has been received.

Thus the commonly accepted view can be replaced by the following formulation: cybernetics is the general science of communication. But to refer to communication is consciously or otherwise to refer to distinguishable states of information inputs and outputs and/or to information being processed within some relatively isolated system. This being so, the widely accepted view can be reformulated thus: cybernetics is the general science of informed and informing systems, and in particular, information systems. This approximates to the statement already made that "the subject matter of cybernetics involves only relatively isolated systems, in particular, etc." — but it seems much more narrow. Yet the still brief history of cybernetics shows that it is precisely cybernetics which has first consciously begun to use the notion of feedback coupling. It must be noted that not every feedback coupling is an information coupling (as will be seen in the chapter on economic models). Should we therefore conclude that cybernetics is not confined to the study of informed and informing systems, but investigates other relatively isolated systems as well, in particular informed and informing systems?

In my opinion, the subject matter of cybernetics is strictly confined to relatively isolated systems. But cybernetics certainly cannot be defined as the science dealing with relatively isolated systems; such a definition would be too broad (cf. *Ref.* A. 1, G. 7, G. 10, L. 2, L. 3, SKL).

Perhaps a bolder thesis? Why, it might be asked, we could not discard the reservation "in particular the informed and informing systems", and simply state that cybernetics is a general theory of relatively isolated systems? Judging from the literature of cybernetics to date, that would be a unilateral extension of frontier. And, moreover, it would involve the occupation of what is not, perhaps, a "no-man's-land", for it seems that a general concept of relatively isolated system is needed not only in cybernetics, but also in two sections of logic — logic of eliminative induction and logic of analogies (cf. *Ref.* G. 5 and G. 9).

3.3. Comparative remarks on basic concepts

Let us now bear in mind what has been said about the subject matter of cybernetics and the division into kinds of fragments of matter, but let us forget the contents of the first two chapters, devoted to basic concepts. And let us then try to guess what must be the general characteristics of the basic concepts of cybernetics.

Since there exist relatively isolated systems which are constructed of most varied kinds of matter (and often, so to speak, of mixed kinds of matter), the basic concepts of cybernetics must be marked by a high degree of abstraction. One may guess, and rightly, that the basic concepts of cybernetics must be so formulated that the *direct* object of study may not be concrete reality, extremely complicated and highly varied, but the somewhat far-fetched idealization.

On a somewhat facetious note, we might say
that cybernetics is concerned with animal, with
man and with telephone alike, but only in the
sense that elementary geometry is concerned with
wood carving, stone splitting and metal cutting.
To put the same idea more "seriously": cybernetics, in a highly abstract way, investigates what
is common to communication processes occurring
both in the human nervous system and in the tele-
graph line; in an analogous way, elementary geometry (or rather: stereometry) investigates the
cutting of solids into component solids, disregarding
the fact that the concept of solid is an idealization
of an approximately rigid body, and irrespective
of whether that body is wood, stone or metal.

The applicability of the basic concepts of cybernetics to highly varied matter is nothing new or
exceptional in the exact sciences. The same property is inherent in the basic concepts of arithmetic, elementary geometry (as mentioned above),
and the theory of probability. The process of count-
ing can refer to stones, animals, men or social organizations. The calculus of probability can be applied
equally well to the disintegration of atomic sub-
stances, to the chances of survival in men or animals, and to the lottery drawing of bonds.

3.4. Theories of relatively isolated systems

We have been introduced to the basic concepts
of cybernetics and have agreed on the subject
matter of that discipline — all this in a very super-

ficial and one-sided way. Now that the tools and the object to be worked on are more or less known, a few words must be devoted to the manner in which those tools are used (the methods of cybernetics) and what has been done with them thus far (the results achieved by cybernetical research).

It might seem at first glance that there is a *general* theory of relatively isolated systems built, as people used to say in former times, *more geometrico*, i. e., as an axiomatic system. But no such theory has been formulated thus far. Nor is that surprising, for such a theory would be the foundation of cybernetics, and it is only a house that is built from its foundations. In the case of a scientific discipline, the foundations are usually laid at a somewhat late stage. Some mathematicians (e. g., Prof. Stanisław Mazur and Dr Antoni Kosiński) object that the formulation of a general theory of relatively isolated systems would not be worth while, since it would have to be too general to include what are called strong theorems. This book, however, is not the proper place to go into that question.

It would be possible to conceive of less general theories of relatively isolated systems, also in an axiomatic form. The objections referred to above would not be applicable to such theories provided that these theories were considerably specialized. Unfortunately, no definite achievements can be recorded here, either. The only exception is the theory of zero-one prospective reliable systems, already comprehensively worked out.

Notable achievements must thus be sought not in a precise formulation of the foundations of cybernetics, but rather in the applications of none too precise basic concepts. Three methods of the application of cybernetics in that respect can be distinguished so far: (1) analysis, (2) synthesis, (3) construction of models. Each of these will now be briefly described.

3.5. Analysis

The analysis of *existing* relatively isolated systems (such as organisms, machines, social organizations) consists in distinguishing within a given system its component parts which in turn are relatively isolated systems, and in studying such couplings which account for the fact that those parts form together a single object.

Understood in this way, such analysis is nothing new to a biologist, a technician, and, perhaps, a sociologist or an economist. And yet it seems that the application of the basic concepts of cybernetics will help any research worker both to perform the analysis and to formulate results lucidly.

3.6. Synthesis

Typical synthesis occurs as follows:
(1) The task is to build a relatively isolated system satisfying certain specified conditions.

(2) A certain range of simple relatively isolated systems is given.

(3) The planned system is built by coupling systems belonging to the given range of simple systems.

(4) The task sometimes proves insoluble. Then, a compromise solution is sought: the task is reformulated so as to become (probably) solvable; for this purpose either the initial conditions are weakened or the given range of simple systems is augmented.

(5) There are often more solutions than one. In such a case, all the solutions are usually studied and the optimum one chosen (from the point of view of costs, efficiency, speed of operation, etc.).

In practice, it is constructors and organizers who, often unconsciously, have to solve problems of synthesis.

3.7. Construction of models

Since the term *"model"* has many different meanings it seems advisable to define in what sense it will be used in this book.

Let us assume that a certain relatively isolated system is given which will be called "the original". (Usually it is an existing system the functioning of which is somewhat complicated but at least superficially known. The nature of the discipline in question determines whether such a system is inanimate non-technical matter, animate non-technical matter, an individual man, a machine, a factory, a social organization or a national economy.)

By "model", we shall here mean a system which is as little complicated as possible and which functions in a manner analogous to the original. "*Model construction*" will mean the designing or physical construction of a model.

The nature of the original may, as already indicated, vary greatly; the models, if they get beyond the mere design stage, are usually constructed of inanimate technical matter.

The purposes for which models are constructed may vary considerably. First of all, *teaching* purposes, as regards which the future of model construction seems enormous. Further, models can be constructed for *research* purposes: if for any reason the original is not readily accessible the model can be studied instead. It is rather obvious that such a study is apt to give unreliable results, and yet in some cases it has, as a method of research, led to valuable discoveries. Finally, model construction is sometimes used for the purposes of *automation*, since a model of the work being done by a worker, a dispatcher or an inspector is the first step towards building an automaton.

(Cf. *Ref.* D. 1, R. 1, T. 1)

3.8. The rôle of mathematics and logic

In cybernetical research (the building of fundamental theories, analysis, synthesis, and model construction) an important rôle is played by the various branches of logic and mathematics, above all:

(1) mathematical logic,
(2) abstract algebras,
(3) functional analysis,
(4) differential equations,
(5) calculus of probability,
(6) mathematical statistics,
(7) theory of games.

So great is the rôle of logic (sentential calculi, lattice theory, Boolean algebra, quantification) that in the opinion of Norbert Wiener, the founder of cybernetics (cf. *Ref.* W. 2), this science would never have emerged at all without mathematical logic.

Later on in the present brief review of cybernetics, we shall restrict our discussion to model construction. Many a logical and/or mathematical theory plays a vital part in that process, but we shall so discuss model construction as to conceal its logical and mathematical aspects.

4. BIOLOGICAL MODELS

4.0. Introductory remarks

The models to be now, though very superficially, discussed are divided into four classes:

(1) biological models,
(2) praxiological models,
(3) logical models,
(4) economic models.

Obviously, this classification is very imperfect and has many gaps, a fact which unavoidably leads to arbitrary decisions in qualifying certain models as belonging to some particular class.

A large branch of cybernetic research consists in building such technical systems (based on mechanical principles, on relays, ferrites, tubes or semi--conductors) which — partially and one-sidedly — imitate manifestations of life. At first glance, such research seems futile, the very endeavour, phantastic. We must, however, bear in mind that from our point of view every living organism is a relatively isolated system built up from simpler relatively isolated systems coupled parallely and/or serially and/or feedback-coupled (cf. *Ref.* C. 1, D. 1, G. 8, G. 10, S. 3).

The branch of cybernetic research in question must now be more closely described, and that in relation to two aspects:

(1) what systems of animate matter are to be represented by technical models, and

(2) what phenomena found in systems of animate matter are to be imitated by technical models.

On the first point — technical models are being built to imitate: (a) an animal in its environment, (b) an animal isolated from its environment, (c) an isolated organ or some other part of an animal. The models mentioned under (b) are usually referred to as synthetic animals.

In reply to the second question, reference must be made to:

(1) homeostasis, i. e., the ability of an organism to maintain an approximate internal equilibrium;

(2) differentiating reactions;

(3) purposive behaviour, in particular the finding of the best environmental conditions (and that in a changing environment);

(4) learning, i. e., formation of habits, by the method of trial and error or of conditioned response;

(5) teaching, i. e., transfer of habits to another individual;

(6) multiplying (including the transfer of "abili ties" in an unchanged or even extended scope).

Ad (1). Homeostasis is achieved in models by means of negative feedback (see Chapter 2). Further, homeostasis is achieved in this way not only in models, but also in the originals (in our sense of the word), i. e., in animals and in men. For instance, it is well known that the self-regulation of the temperature of the system, or the maintaining of arterial blood pressure on more or less the same level is achieved by means of feedback.

Ad (2) and (3). Dr Grey Walter of the Burden Neurological Institute in Bristol has built a number of models in the form of three-wheeled carts which give differentiating reactions and find the best environmental conditions (cf. *Ref.* L. 3).

Ad (4). Learning by the trial and error method is such that at first the response of the original (or model) to the given stimulus is non-purposive to a high degree of probability, but as the number of trials increases the probability of a purposive response also rises, to reach a high level if a sufficient number of trials are undertaken (cf. *Ref.* S. 3).

Ad (4). There exists an opinion — which is both widespread and in error — that feedback is indispensable for modelling a conditioned response. In the next section, a very simple model will be shown which imitates conditioned response without having recourse to feedback. This dates from 1952 and was designed in Poland (cf. *Ref.* GBM and G. 3).

Ad (5). In a later section, a certain model of the learning process will be discussed in some detail. This, too, was designed in Poland, in the year 1957.

Ad (6). Cybernetics has contributed many valuable theoretical data to the very interesting problem of multiplication and heredity. But experiments concerning this aspect were much earlier: in Poland, it was Professor Jan Dembowski who was interested in them (and in the modelling of animals in general) before World War II (cf. *Ref.* D. 1).

4.1. The simplest model of conditioned response

If we have at our disposal but a scanty assortment of zero-one systems (see Chapter 2), we can build a model which imitates conditioned responses, and can do so by coupling elementary zero-one systems serially and parallely, even without resorting to feedback coupling.

A model (probably the simplest possible) imitating conditioned responses will now be built by way of example. For that purpose we couple only:

(a) one alternative system;

(b) two conjunction systems;

(c) two duplicating (bi-replicating) systems;

(d) one delay system.

The model is shown at Fig. 4.1.0. It is clear that we have to do with a zero-one system with two inputs (one principal input and one auxiliary input) and one output. We shall now conduct a number of experiments on that model, it being assumed that we are the first to make any experiments on it.

Γ Stimulus acts on the auxiliary input (i. e., the distinguishable state of that input is 1) but there is no stimulus at the principal input (i. e., the distinguishable state of that input is *0*). What will be the reaction of the sole output of the model?

FIGURE 4.1.0

A learning system

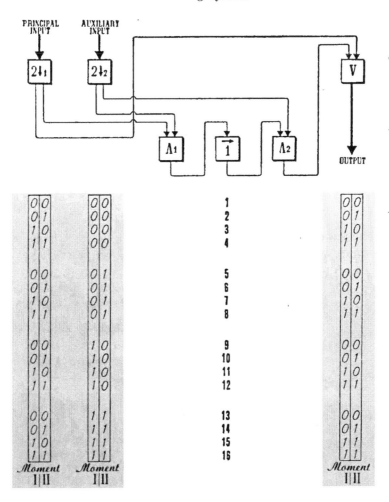

(1) The stimulus at the auxiliary input will be duplicated by the replicating system $2\downarrow_2$ (see Fig. 4.1.0), and, by way of the throughputs

$$\begin{pmatrix} 2\downarrow_2 \\ \wedge_1 \end{pmatrix}, \quad \begin{pmatrix} 2\downarrow_2 \\ \wedge_2 \end{pmatrix},$$

will appear at one input of the conjunction system $\wedge_1$ and at one input of the conjunction system $\wedge_2$.

(2) The distinguishable state *0* at the principal input will be duplicated by the replicating system $2\downarrow_1$ (see Fig. 4.1.0), and, by way of the throughputs

$$\begin{pmatrix} 2\downarrow_1 \\ \vee \end{pmatrix}, \quad \begin{pmatrix} 2\downarrow_1 \\ \wedge_1 \end{pmatrix},$$

will appear at one input of the alternative system $\vee$ and at the second input of the conjunction system $\wedge_1$.

(3) By assumption ("maidenhood" of the system), at the penultimate moment, the state at the input of the delay system $\vec{1}$ (see Fig. 4.1.0) was not *1* (= there was no stimulus).

(4) It follows from (3) that at Moment I the distinguishable state of the throughput

$$\begin{pmatrix} \vec{1} \\ \wedge_2 \end{pmatrix}$$

is *0*.

(5) It follows from (1) and (4) that at one input of the conjunction system $\wedge_2$ the distinguishable state is *1*, and at the other, the distinguishable state is *0*.

(6) It follows from (5) that the distinguishable state of the throughput

$$\begin{pmatrix} \wedge_2 \\ \vee \end{pmatrix}$$

is 0.

(7) It follows from (2) and (6) that the distinguishable state at both inputs of the alternative system $\vee$ is 0.

(8) It follows from (7) that the distinguishable state at the output of the model (see Fig. 4.1.0) is 0.

EXPERIMENT II, AT MOMENT II

The stimulus acts on the principal input of the model (see Fig. 4.1.0 as before), i. e., the distinguishable state of the principal input is 1, and that of the auxiliary input is 0. What will be the reaction of the sole output of the model?

(1) The stimulus at the principal input ($=$ the distinguishable state 1) will be duplicated by the replicating system $2\downarrow_1$, and, by way of the throughputs

$$\begin{pmatrix} 2\downarrow_1 \\ \vee \end{pmatrix}, \quad \begin{pmatrix} 2\downarrow_1 \\ \wedge_1 \end{pmatrix},$$

will appear at one input of the alternative system $\vee$ and at one input of the conjunction system $\wedge_1$.

(2) The distinguishable state 0 at the auxiliary input will be duplicated by the replicating system $2\downarrow_2$, and, by way of the throughputs

$$\begin{pmatrix} 2\downarrow_2 \\ \wedge_1 \end{pmatrix}, \quad \begin{pmatrix} 2\downarrow_2 \\ \wedge_2 \end{pmatrix},$$

will appear at one input of the conjunction system $\wedge_1$ and at one input of the conjunction system $\wedge_2$.

(3) It follows from (1) and (2) of the previous experiment, made at Moment I, that the distinguishable state of the input of the delay system $\overrightarrow{1}$ at Moment I was *0*.

(4) It follows from (3) that the distinguishable state of the output of the delay system $\overrightarrow{1}$, i. e., of the throughput

$$\begin{pmatrix} \overrightarrow{1} \\ \wedge_2 \end{pmatrix},$$

at Moment II, is *0*.

(5) It follows from (2) and (4) that the distinguishable state of both inputs of the conjunction system $\wedge_2$ is *0*.

(6) It follows from (5) that the distinguishable state of the throughput

is *0*. $$\begin{pmatrix} \wedge_2 \\ \vee \end{pmatrix}$$

(7) It follows from (1) and (6) that the distinguishable state of one input of the alternative system $\vee$ is *0*, and that that of the other input is *1*.

(8) It follows from (7) that the distinguishable state of the output of the model is *1*.

The results of Experiments I and II can be thus recorded:

TABLE 4.1.1

Inputs		Output
principal	auxiliary	
0	*1*	*0*
1	*0*	*1*

It might be said in brief that the experiments (in fact replaced by deduction!) have shown that the output reacts to the stimulus acting on the principal input, but does not react to the stimulus acting on the auxiliary input. But we do not stop at this point; we continue our investigations.

EXPERIMENT III, AT MOMENT III

This time the stimulus acts both on the principal input and the auxiliary input, so that the distinguishable state of both inputs of the model is 1.

(1) The stimulus at the principal input is duplicated by the replicating system $2\downarrow_1$, and, by way of the throughputs

$$\binom{2\downarrow_1}{\vee}, \quad \binom{2\downarrow_1}{\wedge_1},$$

the distinguishable state 1 appears at one input of the alternative system $\vee$ and at one input of the conjunction system $\wedge_1$.

(2) The stimulus at the auxiliary input is duplicated by the replicating system $2\downarrow_2$, and, by way of the throughputs

$$\binom{2\downarrow_2}{\wedge_1}, \quad \binom{2\downarrow_2}{\wedge_2},$$

the distinguishable state 1 appears at the second input of the conjunction system $\wedge_1$ and at one input of the conjunction system $\wedge_2$.

(3) It follows from (1) and (2) of Experiment II, made at Moment II, that the distinguishable state

of the input of the delay system $\overrightarrow{1}$ at Moment II was *0*.

(4) It follows from (3), at Moment III, that the distinguishable state of the output of the delay system $\overrightarrow{1}$, i. e., of the throughput

$$\begin{pmatrix} \overrightarrow{1} \\ \wedge_2 \end{pmatrix},$$

is *0*.

(5) It follows from (2) and (4) that the distinguishable state of one input of the conjunction system $\wedge_2$ is *1*, and that that of the other input of that system is *0*.

(6) It follows from (5) that the distinguishable state of the throughput

$$\begin{pmatrix} \wedge_2 \\ \vee \end{pmatrix}$$

is *0*.

(7) It follows from (1) and (6) that the distinguish able state of one input of the alternative system $\vee$ is *1*, and that that of the other input of that system is *0*.

(8) It follows from (7) that the distinguishable state of the output of the model is *1*.

In Experiment III, the stimulus acted both on the principal and on the auxiliary inputs, and reaction was observed at the output. Still one more experiment will be made; at the first glance it will appear to be a mere repetition of Experiment I.

EXPERIMENT IV, AT MOMENT IV

The stimulus acts on the auxiliary input (i. e., the distinguishable state of the auxiliary input is *1*), but does not act on the principal input (i. e., the distinguishable state of the principal input is *0*). What will be the reaction of the sole output of the model?

(1) The stimulus at the auxiliary input will be duplicated by the replicating system $2\downarrow_2$ (see Fig. 4.1.0 as before), and, by way of the throughputs

$$\begin{pmatrix} 2\downarrow_2 \\ \wedge_1 \end{pmatrix}, \quad \begin{pmatrix} 2\downarrow_2 \\ \wedge_2 \end{pmatrix},$$

the distinguishable state *1* will appear at one input of the conjunction system $\wedge_1$ and at one input of the conjunction system $\wedge_2$.

(2) The distinguishable state *0* at the principal input will be duplicated by the replicating system $2\downarrow_1$ (see Fig. 4.1.0), and, by way of the throughputs

$$\begin{pmatrix} 2\downarrow_1 \\ \vee \end{pmatrix}, \quad \begin{pmatrix} 2\downarrow_1 \\ \wedge_1 \end{pmatrix},$$

will appear at one input of the alternative system $\vee$ and at the second input of the conjunction system $\wedge_1$.

(3) It follows from (1) and (2) of Experiment III, at Moment III, that the distinguishable state of the input of the delay system $\overrightarrow{1}$, i. e., of the throughput

$$\begin{pmatrix} \wedge_1 \\ \overrightarrow{1} \end{pmatrix}$$

at Moment III, was *1*.

(4) It follows from (3), that at present, i. e., at Moment IV, the distinguishable state of the output of the delay system $\vec{1}$, i. e., of the throughput

$$\left(\begin{array}{c} \vec{1} \\ \wedge_2 \end{array}\right),$$

is *1*.

(5) It follows from (1) and (4) that the distinguishable state of both inputs of the conjunction system $\wedge_2$ is *1*.

(6) It follows from (5) that the distinguishable state of the throughput

$$\left(\begin{array}{c} \wedge_2 \\ \vee \end{array}\right)$$

is *1*.

(7) It follows from (2) and (6) that the distinguishable state of one input of the alternative system $\vee$ is *0*, and that that of the other input of that system is *1*.

(8) It follows from (7) that the distinguishable state of the output of the alternative system $\vee$, i. e., of the output of the model, is *1*.

Thus, this time, contrary to the result of Experiment I, the distinguishable state of the output is *1*, while the distinguishable states of the inputs are: *0* for the principal input, *1* for the auxiliary input.

The results of the four Experiments are summed up below in Table 4.1.2.

TABLE 4.1.2

Moment	Distinguishable state of the		Distinguishable state of the output
	principal input	auxiliary input	
I	0	1	0
II	1	0	1
III	1	1	1
IV	0	1	1

Careful analysis of the diagram of the model (Fig. 4.1.0), or careful analysis of the four Experiments (in fact replaced here by four deductions), or even analysis of the results presented in Table 4. 1. 2, enables us to conclude that the result of a given experiment with our model depends not only on the state of inputs in the given moment, but also on the state of such inputs in the penultimate moment.

An analogy was the case with the well-known Pavlov experiments. It is perhaps superfluous to explain that the distinguishable state *1* of the principal input corresponds to the smell or sight of the food, the distinguishable state *1* of the auxiliary input, to the ringing of the bell, and the distinguishable state *1* of the output, to the increased salivation in the dog.

The "learning" model shown at Fig. 4.1.0 is "both a genius and an imbecile", a "genius", because a *single* coincidence of stimuli at the principal and the auxiliary inputs is sufficient to develop in it a conditioned response; an "imbecile", because it reveals such conditioned response *only once* after each training. Thus we have to do with a system which "learns" and "forgets" very easily.

The four experiments do not, of course, exhaust all the possible combinations of states of the two inputs in two consecutive moments. The number of such combinations is easily shown to be 16. The lower part of Fig. 4.1.0 shows all those possible combinations of states of the two inputs in two consecutive moments, and the corresponding states of the output.

(Cf. *Ref.* G. 3, G. 8, G. 9, GBM)

4.2. Discussion

A number of objections might be raised against the extremely simple model imitating the conditioned response as presented above. The most important of such objections would perhaps be:

(1) There are no known organisms which would learn and forget so rapidly as does the model in question.

(2) During the training, the stimulus at the auxiliary input should precede the stimulus at the principal input, since that is the order in the case of training animals.

(3) The model does not show any differentiation of the stimulus. It is known that a dog on which experiments are being made, when it has once associated the smell of the food with the drawing of a circle, at first also reacts with increased salivation to the drawing of an ellipse.

All these objections can be confidently waived. To build models more complicated and yet consisting only of zero-one elements presents no difficulty; to such models none of the objections raised above would be applicable.

It must be frankly admitted that the designing of "learning" zero-one models, even extremely complicated ones, requires today no technical or engineering ingenuity; such ingenuity has by now been completely replaced by algebraic and logical calculus (thus has mathematical logic taken "vengeance" on all those who used to ridicule it as useless and inapplicable to any practical purposes).

But our opponent may say that even an improvement on our model is not satisfactory since it is a *reliable* system, whereas the results of experimental physiological research lead him to believe that the conditioned response is always probabilistic in nature, and hence our model should be an *unreliable* system.

The reply is — that such a system can also be built.*

But our opponent may still remain unconvinced. Even if he is not a vitalist, he may say that when conditioned response is concerned there is no point in building *any* model (in our sense of the word), since every model is built of inanimate matter, whereas every original is built of animate matter, and there is a qualitative difference between these

---

* Several years ago, Professor Stefan Manczarski (Warsaw) built a probabilistic model imitating the conditioned response.

two kinds of matter. Consequently, there is no point whatever in building and improving models able to imitate the conditioned response.

But even to this we reply:

(1) Purposiveness is a relative attribute. Any action is, or is not, purposive with respect to some end. In the case of the modelling of the conditioned response, the following ends may be involved:

(a) teaching purposes;

(b) specialized gnoseological purposes;

(c) technical purposes (automation);

(d) philosophical purposes.

(2) In the preceding chapter, we have emphasized the undoubted value of modelling for teaching purposes, and the dubiety of the gnoseological value. Hardly anything remains to be added on those points.

(3) The rôle of modelling as a stage preparatory to building automata has also been mentioned. But one point can be added here: the modelling of the conditioned response may, although probably in a remote future, lead to a new concept of the automaton, or rather a hyperautomaton, that is, a device which would control a machine and would adjust itself to the individual properties (even temporary) of the machine in question.

(4) Without questioning the existence of qualitative difference as between inanimate and animate matter, we must affirm that it is precisely from the philosophical point of view that the modelling of the conditioned response in inanimate technical

matter is of particular interest. Biological phenomena, and especially the conditioned response,
ought to be modelled with great care and in great
detail, since only such models can provide new
information concerning the boundary between animate and inanimate (technical) matter.

4.3. Modelling of teaching processes

Let us consider a certain object which satisfies
all the following conditions:

(1) It consists of two living beings.

(2) The first of these beings can develop conditioned responses.

(3) The other being has the same property.

(4) If the first being is trained in a way sufficient for it to develop a conditioned response, then
it is not necessary to train the other being, since
the first directly transmits to it the conditioned
response it has just developed.

Our task now is to build the simplest possible
model of such an object. To do so we proceed as
follows:

(1) We build *two copies* of the "learning" model,
already known to us (Fig. 4.1.0).

(2) One of these two systems is expanded, so that
we add to it one duplicating system $2\downarrow_3$; we do
so (see Fig. 4.1.0) by

(a) breaking the throughput

$$\begin{pmatrix} \wedge_2 \\ \vee \end{pmatrix},$$

(b) building two new throughputs

$$\begin{pmatrix} \wedge_2 \\ 2\downarrow_3 \end{pmatrix}, \quad \begin{pmatrix} 2\downarrow_3 \\ \vee \end{pmatrix}.$$

In this way, we obtain a new system (see Fig. 4.3.0, upper part) which has two outputs instead of one, viz., the outward output of the replicating system $2\downarrow_3$, and the outward output of the alternative system $\vee_1$. This new system will be called the "teaching system", and its outward output from the replicating system $2\downarrow_3$, the "communication effector" of the teaching system. It can be easily verified that this "surgical operation" has not deprived the teaching system of the ability to develop conditioned responses, but has made it possible for it "to express itself outwards".

(3) The other copy of the "learning" model is also expanded, but in a different way. We replace the duplicating system $2\downarrow_3$ (see Fig. 4.1.0) by the triplicating system $3\downarrow$ (see Fig. 4.3.0, lower part), and moreover we add one conjunction system $\wedge_5$ and one alternative system $\vee_3$ (see Fig. 4.3.0 again). This is done as follows:

(a) we build the throughput

$$\begin{pmatrix} 3\downarrow \\ \wedge_5 \end{pmatrix}, \quad \text{and}$$

(b) the throughputs

$$\begin{pmatrix} \vee_2 \\ \vee_3 \end{pmatrix}, \quad \begin{pmatrix} \wedge_5 \\ \vee_3 \end{pmatrix}.$$

In this way, we obtain a new system (Fig. 4.3.0, lower part) which, in addition to the two inputs

it had before (the principal and the auxiliary),
now has a third input, viz., to the conjunction
system $\wedge_5$. This new system will be called the
"taught system", and the new input, the "communication receptor" of the taught system. It can
easily be seen that this "surgical operation" has
not deprived the system under discussion of its
ability to develop conditioned responses, and has
made it possible for it to, so to speak, "receive
signals from outside".

(4) We couple the teaching system serially with
the taught system by building the throughput

$$\begin{pmatrix} 2\downarrow_3 \\ \wedge_5 \end{pmatrix}$$

(see Fig. 4.3.0, both parts).

Let us see now what has been the result of our
somewhat complicated assembly operations, and
whether the object described above (a coupled
pair of living beings) really has received its model.

Suppose that at Moment I the distinguishable
state of the input of the delay system $\vec{1}_1$ and of
the input of the delay system $\vec{1}_2$ is 0. It is clear
that if in such conditions at Moment I, the
distinguishable state of both the principal and the
auxiliary input of the teaching system is 1, then it
is sufficient for the auxiliary input of the teaching
system and the taught system to have the distinguishable state 1 at Moment II in order to have,
in the same Moment II, the distinguishable state 1
of its output. To put it perhaps less exactly but

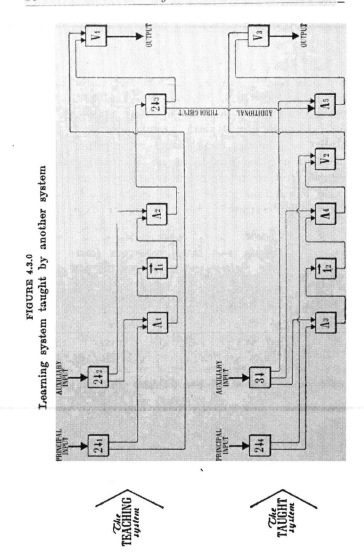

FIGURE 4.3.0

Learning system taught by another system

somewhat more picturesquely: the training of the teaching system is sufficient for the taught system to develop symptoms of a conditioned response without having been trained itself.

It is also worth while to analyse the reactions of our model (a teaching system coupled serially with a taught system — Fig. 4.3.0, both parts) for all the possible distinguishable states of four inputs

TABLE 4.3.1

The distinguishable state of the output of the taught system at Moments I and II

THE TAUGHT SYSTEM

		PRINCIPAL INPUT															
		00	01	10	11	00	01	10	11	00	01	10	11	00	01	10	11
		00				01				10				11				(AUXILIARY INPUT)
00	00	00	01	10	11	00	01	10	11	00	01	10	11	00	01	11	11
01		00	01	10	11	00	01	10	11	00	01	10	11	00	01	11	11
10		00	01	10	11	00	01	10	11	00	01	10	11	00	01	11	11
11		00	01	10	11	00	01	10	11	00	01	10	11	00	01	11	11
00	01	00	01	10	11	00	01	10	11	00	01	10	11	00	01	11	11
01		00	01	10	11	00	01	10	11	00	01	10	11	00	01	11	11
10		00	01	10	11	00	01	10	11	00	01	10	11	00	01	11	11
11		00	01	10	11	00	01	10	11	00	01	10	11	00	01	11	11
00	10	00	01	10	11	00	01	10	11	00	01	10	11	00	01	11	11
01		00	01	10	11	00	01	10	11	00	01	10	11	00	01	11	11
10		00	01	10	11	00	01	10	11	00	01	10	11	00	01	11	11
11		00	01	10	11	00	01	10	11	00	01	10	11	00	01	11	11
00	11	00	01	10	11	00	01	10	11	00	01	10	11	00	01	11	11
01		00	01	10	11	00	01	10	11	00	01	10	11	00	01	11	11
10		00	01	10	11	01	01	11	11	00	01	10	11	01	01	11	11
11		00	01	10	11	01	01	11	11	00	01	10	11	01	01	11	11

THE TEACHING SYSTEM (left axis: PRINCIPAL INPUT, AUXILIARY INPUT)

In each two-digit group the first (left-hand) digit shows the distinguishable state at Moment I, and the second (right-hand) digit, that at Moment II.

(two principal and two auxiliary) during two consecutive moments. Here, this analysis will be confined to the study of the reactions of the output of the taught system (see Table 4.3.1).

In one moment, each of the four inputs takes on only one of the *two* distinguishable states: either *0*, or *1*. Consequently, in two consecutive moments, each input takes on only one of the following four distinguishable states:

00, 01, 10, 11.

This means that for each pair of inputs in two consecutive moments we must take into consideration

$$4 \cdot 4 = 16$$

combinations, the corresponding figure for all the four inputs being

$$16 \cdot 16 = 256$$

combinations. Each of the 256 squares in Table 4.3.1 represents the distinguishable state of the output of the taught system at Moment I (left digit), and at Moment II (right digit). The Table must have a fairly uniform appearance since each of these 256 squares must be filled with one of the *only* four combinations of digits:

00, 01, 10, 11.

If in that Table a whole column from top to bottom is filled with one and the same combination of digits, then, and only then, we have to do with distinguishable states of the principal and the auxiliary input of the taught system for which the states of input of the teaching system do not influence the distinguishable state of the output of the taught system. Of the 16 columns in Table 4.3.1, thirteen columns reveal such uniformity. The remaining three are for us of particular interest since they illustrate the influence exerted by the teaching system on the taught system. They are columns corresponding to the following states of the *taught* system:

$$\text{Input} \begin{cases} \text{principal:} & \textit{00, 10, 00,} \\ \text{auxiliary:} & \textit{01, 01, 11.} \end{cases}$$

It must also be explained that in Table 4.3.1 those digits "*1*" which refer to the result of a conditioned response in the taught system are underlined once, and those digits "*1*" which refer to the appearance of influence exerted by the teaching system on the taught system are underlined twice.

(Cf. *Ref.* G. 8)

4.4. Concluding remarks

It seems advisable to add here the following information:

(1) There is no difficulty in modelling such a pair of living beings in which teaching, or rather transmission of experience, is reciprocal. In such a case we have to do with a pair of systems each of which

is both a teaching and a taught system; they must be *feedback*-coupled, and not merely serially coupled, as in our very simple example above.

(2) It is also possible to build such models as, after a period of training, permanently acquire a given conditioned response. It would be difficult to assert, without first having formulated a precise termininology (which exigencies of space prevent us from doing here), that in each such system the acquired conditioned response becomes an unconditioned response. Each such system must be self--coupled (see Section 2.11).

(3) The concept of conditioned response (called time function) can be so defined that it ceases to be a biological concept and becomes a cybernetic one. Such a generalization reveals that the conditioned response is observable not only in the *B*, *C* and *E* kinds of matter (see Section 3.1), but also in very highly organized forms of matter, viz., in human communities. This seems to be a very promising subject for further investigation.

5. PRAXIOLOGICAL MODELS

5.0. Introductory remarks

By praxiological models we mean all models, and only such models, which illustrate the interaction between the agent (individual or collective) and the environment or the interaction between individual agents (or else the one-sided action by such agents). The concepts of cybernetics are useful in the construction of such models, the most valuable being those of indirect serial coupling, and, in particular, indirect feedback coupling.

The graphical method of representing systems and couplings will be enriched with a number of new symbols:

(1) R — (or R_1, R_2, ...) — receptor,
(2) E — (or E_1, E_2, ...) — effector,
(3) N — (or N_1, N_2, ...) — central nervous system,
(4) IR — observation instrument (see Section 5.1),
(5) IE — performance instrument (see Section 5.1),
(6) IN — intellectual instrument (see Section 5.1),
(7) Obi — object of action.

All drawings in this Chapter will, for the purpose of emphasizing the rôle of memory in the functioning of the system N, show the central nervous system N always as a self-coupled system.

(Cf. *Ref.* BGS and K. 1)

5.1. The agent and the object of action

When we act on an object by means of an effector (in a model, by means of one effector, in reality by means of a large set of effectors) we almost instantaneously receive information about that object by way of a receptor (in a model, by way of one receptor, in reality by means of a large set of receptors). For instance, we put a nut on a screw, turn the nut with the fingers, and with our eyes observe what we are doing (see Fig. 5.1.0). This is how both men and animals behave. Thus, a general model of the contacts between the agent and the object of action can be built as a relatively isolated system obtained by means of an *indirect* feedback coupling of two systems — the agent and the object of action — or else, of four systems (see Fig. 5.1.0):

$$R, N, E, Obi,$$

which are coupled by means of the throughputs:

$$\binom{R}{N}_{;} \quad \binom{N}{E}, \quad \binom{E}{Obi}, \quad \binom{Obi}{R}_{;}$$

If the object of action (*Obi*) remains passive, then the throughput

$$\binom{Obi}{R}$$

is an information input to the system R.

Each of the four systems:

$$R, N, E, Obi$$

has not only inputs and outputs involved in the serial couplings referred to above, but also inputs

from the outside and outputs to the outside of the
model. This indicates unreliability of action in the
various systems of the model (unreliability of infor-

FIGURE 5.1.0
Human being, or animal

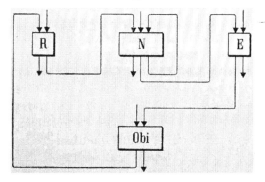

mation transmitted by the system R, unreliability
of decisions taken by the system N, and unreliability of actions undertaken by the system E), and also
a possibility of "unforeseen" behaviour by the
system Obi.

As already mentioned, the model shown at Fig.
5.1.0 is applicable to both men and animals. But
as regards men, even primitives, the model is over-
simplified, since it disregards two factors inseparable
from human behaviour: tools (more generally:
instruments) and language. Later on, we shall
deal with models which take into consideration
instruments of all kinds, and our investigations
will be rounded off by an outline of a model which
takes into account the rôle of language in human
behaviour.

In our — extremely schematic — analysis instruments are classified into three groups:

(1) performance instruments,

(2) observation instruments,

(3) intellectual instruments.

This requires certain terminological explanations.

Ad (1). By a *"performance instrument"* we mean every instrument, and only such, which replaces or amplifies an *effector*, or makes it possible for that effector to undertake some modified action.

Ad (2). By an *"observation instrument"* we mean every instrument, and only such, which replaces or amplifies a *receptor*, or makes it possible for that receptor to undertake some modified action.

Ad (3). By an *"intellectual instrument"* we mean every instrument, and only such, which replaces or amplifies a *central nervous system*, or makes it possible for that central nervous system to undertake some modified action.

The order in which these three kinds of instruments are enumerated above seems roughly to coincide with the order in which they appeared in the history of mankind. The same order will be adhered to in the discussion of more developed praxiological models (cf. *Ref.* BGS).

5.2. Unregulated performance instruments

It is fairly commonly known that a nail is driven into wood not with the fist, but with a hammer. It is also known, though perhaps less commonly, that larger pieces of meat are not torn apart with

hands (and teeth), but pinned with a fork and
cut with a knife. The hammer, the fork and the
knife are simple examples of unregulated instruments which amplify certain effectors. An extremely simplified model of action by means of such
an instrument is shown at Fig. 5.2.0. We have to
do there with (as compared with the previous case)

FIGURE 5.2.0
Human being using an unregulated performance instrument

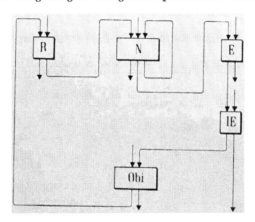

an increased number of indirect serial couplings,
which together give an indirect feedback coupling
built of the throughputs:

$$\binom{R}{N}, \quad \binom{N}{E}, \quad \binom{E}{IE}, \quad \binom{IE}{Obi}, \quad \binom{Obi}{R}.$$

As in the former case, the inputs and outputs
of the systems pertaining to our model, and which
must be taken into account, are those not involved

in the couplings shown above (inputs from the outside and outputs to the outside of the model); this refers in particular to the input from the outside of the model to the system *IE* (indicating new sources of unreliability of action), and the output to the outside of the model from the system *IE* (indicating the possibility of side effects not intended by the agent). It must be emphasized that the input from the outside of the model to the system *IE* may be a source of power, e. g., when the system *IE* is a motor.

The matrix method of showing the couplings of a system was introduced in Section 2.13. That method is particularly valuable when a large num-

TABLE 5.2.1

Input	Output						Total of outputs
	from the outside	*R*	*N*	*E*	*IE*	*Obi*	
to the outside	·	1	1	1	1	1	5
R	1	0	0	0	0	1	2
N	1	1	1	0	0	0	3
E	1	0	1	0	0	0	2
IE	1	0	0	1	0	0	2
Obi	1	0	0	0	1	0	2
Total of inputs	5	2	3	2	2	2	16

ber of systems is coupled, since in such cases the
graphical method (see Section 2.1) fails almost
entirely. The model now under discussion (see Fig.
5.2.0) involves the coupling of only five systems:

$$R, N, E, IE, Obi$$

and can be lucidly presented by the graphical
method; but for didactic reasons it seems advisable to present the zero-one matrix of that model
and to ask the reader to confront the graph (5.2.0)
with the matrix (5.2.1).

5.3. Regulated performance instruments

If a given performance instrument can be regulated, as is, e. g., the monkey-wrench, then the
network of couplings becomes more complicated as
compared with the previous model (see Fig. 5.2.0),
since the new effector E_2, used to regulate the
instrument, must be taken into account, and also
the new receptor R_2, used to control the regulation
of the instrument (see Fig. 5.3.0). The graph includes the following loops which correspond to feedbacks:

(1) the already familiar shortest loop representing the self-coupling of the central nervous system

$$\binom{N}{N};$$

(2) the known loop:

$$\binom{R_1}{N}, \quad \binom{N}{E_1}, \quad \binom{E}{IE}, \quad \binom{IE}{Obi}, \quad \binom{Obi}{R_1};$$

(3) a new loop, much more complicated:

$$\binom{R_1}{N}, \quad \binom{N}{E_2}, \quad \binom{E_2}{IE}, \quad \binom{IE}{R_2},$$

$$\binom{R_2}{N}, \quad \binom{N}{E_1}, \quad \binom{E_1}{IE}, \quad \binom{IE}{Obi}, \quad \binom{Obi}{R_1}.$$

It seems advisable to give an example calculated to help the reader to associate this feedback with what he knows perfectly well from every-day life.

FIGURE 5.3.0

Human being using a regulated performance instrument

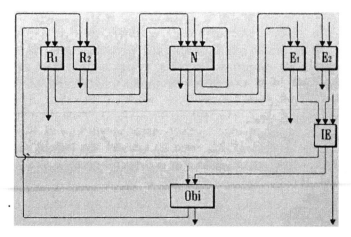

In such example the various systems involved in the model play the following rôles:

(1) *Obi* — the nut,

(2) R_1 — the eyes looking at the nut,

(3) R_2 — the eyes looking at the monkey-wrench,

(4) N — the central nervous system,

(5) E_1 — the hand holding the monkey-wrench,
(6) E_2 — the fingers regulating the monkey-wrench,
(7) IE — the monkey-wrench.

The functioning of the loop (3):

$$\left.\begin{matrix}\begin{pmatrix}Obi\\R_1\end{pmatrix}\\[1em]\begin{pmatrix}R_1\\N\end{pmatrix}\end{matrix}\right\}$$ — view of the nut not yet screwed on,

$$\left.\begin{matrix}\begin{pmatrix}N\\E_2\end{pmatrix}\\[1em]\begin{pmatrix}E_2\\IE\end{pmatrix}\end{matrix}\right\}$$ — monkey-wrench adjusted to the nut,

$$\left.\begin{matrix}\begin{pmatrix}IE\\R_2\end{pmatrix}\\[1em]\begin{pmatrix}R_2\\N\end{pmatrix}\end{matrix}\right\}$$ — view of the monkey-wrench adjusted to the nut,

$$\left.\begin{matrix}\begin{pmatrix}N\\E_1\end{pmatrix}\\[1em]\begin{pmatrix}E_1\\IE\end{pmatrix}\end{matrix}\right\}$$ — the hand turns the monkey-wrench,

$$\begin{pmatrix}IE\\Obi\end{pmatrix}$$ — the monkey-wrench screws on the nut,

$$\left.\begin{matrix}\begin{pmatrix}Obi\\R_1\end{pmatrix}\\[1em]\begin{pmatrix}R_1\\N\end{pmatrix}\end{matrix}\right\}$$ — view of the nut screwed on (or being screwed on).

For practical purposes, it seems advisable to give the matrix of couplings descriptive of the model (see Table 5.3.1).

TABLE 5.3.1

Input	from the outside	R_1	R_2	N	E_1	E_2	IE	Obi	Total of outputs
				Output					
to the outside	·	1	1	1	1	1	1	1	7
R_1	1	0	0	0	0	0	0	1	2
R_2	1	0	0	0	0	0	1	0	2
N	1	1	1	1	0	0	0	0	4
E_1	1	0	0	1	0	0	0	0	2
E_2	1	0	0	1	0	0	0	0	2
IE	1	0	0	0	1	1	0	0	3
Obi	1	0	0	0	0	0	1	0	2
Total of inputs	7	2	2	4	2	2	3	2	24

5.4. Performance instruments coupled with a substitutive effector

I have broken my leg, I cannot put my weight on it, but I can walk with crutches. In such a case, certain effectors are replaced by others acting through a performance effector (here, the crutches).

Let us now suppose that my leg has been put right, yet I can walk only with difficulty; I no longer use the crutches, but I have to lean on a stick. Now the effectors of the restored leg are already working, but have to be helped by the substitutive effectors of the hand and act through a performance instrument — the walking-stick — coupled with them.

The model is shown at Fig. 5.4.0.

FIGURE 5.4.0

Human being using an unregulated performance instrument coupled with a substitutive effector

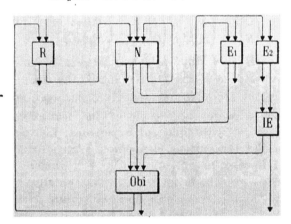

It is built of six elementary systems:

$$R, N, E_1, E_2, IE, Obi.$$

Let us now pay attention to the following feedback couplings:

(1) The self-coupling of the central nervous system:

$$\binom{N}{N};$$

(2) The coupling built of the four throughputs:

$$\binom{R}{N}, \quad \binom{N}{E_1}, \quad \binom{E_1}{Obi}, \quad \binom{Obi}{R};$$

(3) The coupling built of the five throughputs:

$$\binom{R}{N}, \quad \binom{N}{E_2}, \quad \binom{E_2}{IE}, \quad \binom{IE}{Obi}, \quad \binom{Obi}{R};$$

(4) The coupling built of the seven throughputs:

$$\binom{R}{N}, \quad \binom{N}{E_1}, \quad \binom{E_1}{Obi}, \quad \binom{Obi}{R},$$

$$\binom{R}{N}, \quad \binom{N}{E_2}, \quad \binom{E_2}{IE}, \quad \binom{IE}{Obi}, \quad \binom{Obi}{R}.$$

This time, too, it seems reasonable to help the reader in associating each of the feedbacks (2), (3) and (4) with practical experience. Let us suppose that the various elementary systems will play the following rôles:

(1) Obi — the road on which I am walking,

(2) R — the set of tactile and visual receptors which inform me as to how I am walking,

(3) N — my central nervous system,

(4) E_1 — the set of effectors of my newly restored leg,

(5) E_2 — the set of effectors of the hand in which I am holding my stick,

(6) IE — the walking-stick I lean on.

The rôle of the feedback

$$\begin{pmatrix} R \\ N \end{pmatrix}, \quad \begin{pmatrix} N \\ E_1 \end{pmatrix}, \quad \begin{pmatrix} E_1 \\ Obi \end{pmatrix}, \quad \begin{pmatrix} Obi \\ R \end{pmatrix}$$

or the feedback

$$\begin{pmatrix} Obi \\ R \end{pmatrix}, \quad \begin{pmatrix} R \\ N \end{pmatrix}, \quad \begin{pmatrix} N \\ E_1 \end{pmatrix}, \quad \begin{pmatrix} E_1 \\ Obi \end{pmatrix}$$

can be, e. g., as follows (see Fig. 5.4.0): because I feel pain in my newly restored leg, I start walking on the ball of the foot.

The rôle of the feedback

$$\begin{pmatrix} R \\ N \end{pmatrix}, \quad \begin{pmatrix} N \\ E_2 \end{pmatrix}, \quad \begin{pmatrix} E_2 \\ IE \end{pmatrix}, \quad \begin{pmatrix} IE \\ Obi \end{pmatrix}, \quad \begin{pmatrix} Obi \\ R \end{pmatrix}$$

or the feedback

$$\begin{pmatrix} Obi \\ R \end{pmatrix}, \quad \begin{pmatrix} R \\ N \end{pmatrix}, \quad \begin{pmatrix} N \\ E_2 \end{pmatrix}, \quad \begin{pmatrix} E_2 \\ IE \end{pmatrix}, \quad \begin{pmatrix} IE \\ Obi \end{pmatrix}$$

(see Fig. 5.4.0), can be illustrated as follows: my walking-stick, not held quite vertically to the road surface, has slipped; I adjust the position of the walking-stick and now feel that I can safely lean on it since it holds the road surface well.

More complicated is the rôle of the feedback

$$\begin{pmatrix} R \\ N \end{pmatrix}, \quad \begin{pmatrix} N \\ E_1 \end{pmatrix}, \quad \begin{pmatrix} E_1 \\ Obi \end{pmatrix}, \quad \begin{pmatrix} Obi \\ R \end{pmatrix},$$

$$\begin{pmatrix} R \\ N \end{pmatrix}, \quad \begin{pmatrix} N \\ E_2 \end{pmatrix}, \quad \begin{pmatrix} E_2 \\ IE \end{pmatrix}, \quad \begin{pmatrix} IE \\ Obi \end{pmatrix}, \quad \begin{pmatrix} Obi \\ R \end{pmatrix}$$

or the feedback

$$\begin{pmatrix} Obi \\ R \end{pmatrix}, \quad \begin{pmatrix} R \\ N \end{pmatrix}, \quad \begin{pmatrix} N \\ E_1 \end{pmatrix}, \quad \begin{pmatrix} E_1 \\ Obi \end{pmatrix},$$

$$\begin{pmatrix} Obi \\ R \end{pmatrix}, \quad \begin{pmatrix} R \\ N \end{pmatrix}, \quad \begin{pmatrix} N \\ E_2 \end{pmatrix}, \quad \begin{pmatrix} E_2 \\ IE \end{pmatrix}, \quad \begin{pmatrix} IE \\ Obi \end{pmatrix}.$$

The rôle of that somewhat intricate loop can be explained by the following example: I am conscious of pain in the heel of the newly restored leg, so I stand carefully on the balls of my feet; I feel that my leg slips forward, I thrust out my walking-stick and lean heavily on it, which prevents me from falling down.

The feedback (4) can also be shown as

$$\begin{pmatrix} R \\ N \end{pmatrix}, \quad \begin{pmatrix} N \\ E_2 \end{pmatrix}, \quad \begin{pmatrix} E_2 \\ IE \end{pmatrix}, \quad \begin{pmatrix} IE \\ Obi \end{pmatrix}, \quad \begin{pmatrix} Obi \\ R \end{pmatrix},$$

$$\begin{pmatrix} R \\ N \end{pmatrix}, \quad \begin{pmatrix} N \\ E_1 \end{pmatrix}, \quad \begin{pmatrix} E_1 \\ Obi \end{pmatrix}, \quad \begin{pmatrix} Obi \\ R \end{pmatrix}$$

or (without any change in its cyclic order) as

$$\begin{pmatrix} Obi \\ R \end{pmatrix}, \quad \begin{pmatrix} R \\ N \end{pmatrix}, \quad \begin{pmatrix} N \\ E_2 \end{pmatrix}, \quad \begin{pmatrix} E_2 \\ IE \end{pmatrix}, \quad \begin{pmatrix} IE \\ Obi \end{pmatrix},$$

$$\begin{pmatrix} Obi \\ R \end{pmatrix}, \quad \begin{pmatrix} R \\ N \end{pmatrix}, \quad \begin{pmatrix} N \\ E_1 \end{pmatrix}, \quad \begin{pmatrix} E_1 \\ Obi \end{pmatrix}.$$

As an example of its functioning: I feel that the road is not so slippery so I lean on my walking-stick less heavily, my balance is upset, I put my weight firmly on my newly restored leg and regain my balance.

The zero-one matrix of couplings of the model is given at Table 5.4.1.

TABLE 5.4.1

Input	Output							Total of outputs
	from the outside	R	N	E_1	E_2	IE	Obi	
to the outside	.	1	1	1	1	1	1	6
R	1	0	0	0	0	0	1	2
N	1	1	1	0	0	0	0	3
E_1	1	0	1	0	0	0	0	2
E_2	1	0	1	0	0	0	0	2
IE	1	0	0	0	1	0	0	2
Obi	1	0	0	1	0	1	0	3
Total of inputs	6	2	4	2	2	2	2	20

5.5. Unregulated observation instruments

· When a long-sighted man wants to see an object situated too near to him, he usually puts on (suitable) spectacles. When a short-sighted man wants to see a remote object, he too puts on (different) spectacles. The spectacles are good examples of a relatively isolated system which is an unregulated instrument amplifying a receptor.

Fig. 5.5.0 shows a more complicated type of the model familiar from Fig. 5.1.0. In the case of 5.5.0,

we have to do with one self-coupling and one feedback, the latter being built of the following throughputs:

$$\begin{pmatrix} R \\ N \end{pmatrix}, \quad \begin{pmatrix} N \\ E \end{pmatrix}, \quad \begin{pmatrix} E \\ Obi \end{pmatrix}, \quad \begin{pmatrix} Obi \\ IR \end{pmatrix}, \quad \begin{pmatrix} IR \\ R \end{pmatrix}.$$

Attention must be drawn to the fact that each
of the five elementary systems (including the system *IR*), of which the model now under discussion
is built, has one input and one output not involved
in the inner couplings within the model. An extra
input to *IR* indicates that the information about

FIGURE 5.5.0

Human being using an unregulated observation instrument

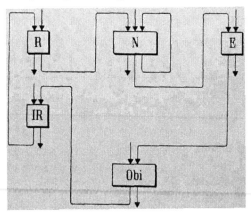

the system *Obi*, reaching the system *R* via the system *IR*, may be distorted within *IR* (if, e. g., the
glasses of the spectacles become misted). An extra
output from *IR* indicates that over and above

conveying information to R, the system IR may also have some side-effects (e. g., the spectacles may chafe behind the ears).

The matrix of couplings of the model is as follows (see Table 5.5.1):

TABLE 5.5.1

Input	Output						Total of outputs
	from the outside	IR	R	N	E	Obi	
to the outside	.	1	1	1	1	1	5
IR	1	0	0	0	0	1	2
R	1	1	0	0	0	0	2
N	1	0	1	1	0	0	3
E	1	0	0	1	0	0	2
Obi	1	0	0	0	1	0	2
Total of inputs	5	2	2	3	2	2	16

5.6. Regulated observation instruments

Microscope and telescope are good examples of regulated instruments amplifying a receptor. In this case, the model is more complicated than that familiar from the preceding Section. The structure of the model offers, of course, a certain analogy to the structure of the model described in Section 5.3 ("Regulated performance instruments"). Fig. 5.6.0 shows the model we are now discussing; it is built of seven systems:

$$IR, R_1, R_2, N, E_1, E_2, Obi.$$

FIGURE 5.6.0

Human being using a regulated observation instrument

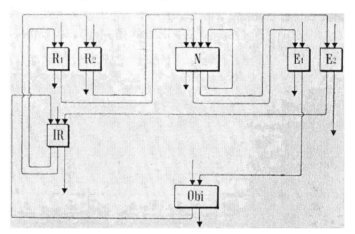

The feedbacks in this model are:

(1) The well known shortest possible loop of the self-coupling of the central nervous system

$$\binom{N}{N};$$

(2) the loop

$$\binom{R_1}{N}, \quad \binom{N}{E_1}, \quad \binom{E_1}{Obi}, \quad \binom{Obi}{IR}, \quad \binom{IR}{R_1},$$

already known from the preceding Section (Fig. 5.5.0);

(3) the more complicated loop

$$\binom{R_1}{N}, \quad \binom{N}{E_2}, \quad \binom{E_2}{IR}, \quad \binom{IR}{R_2},$$

$$\binom{R_2}{N}, \quad \binom{N}{E_1}, \quad \binom{E_1}{Obi}, \quad \binom{Obi}{IR}, \quad \binom{IR}{R_1}.$$

Of course, feedback (3) can be symbolized in a different way, beginning with another through-put but without changing the cyclic order:

$$\binom{Obi}{IR}, \quad \binom{IR}{R_1}, \quad \binom{R_1}{N}, \quad \binom{N}{E_2},$$

$$\binom{E_2}{IR}, \quad \binom{IR}{R_2}, \quad \binom{R_2}{N}, \quad \binom{N}{E_1}, \quad \binom{E_1}{Obi}.$$

An example will now explain to the reader the function of feedback (3); very familiar from practical experience. To the various systems forming the model (Fig. 5.6.0) the following rôles will be allotted:

(1) Obi — the hill which I am to climb and the road leading to it,

(2) IR — field binoculars,

(3) R_1 — my eyes (the set of receptors involved),

(4) R_2 — tactile receptors — my fingers — which adjust the binoculars,

(5) N — my central nervous system,

(6) E_1 — my legs (the set of effectors involved),

(7) E_2 — effectors of my fingers adjusting the binoculars.

The functioning of the loop:

$$\left. \begin{matrix} \binom{Obi}{IR} \\ \\ \binom{IR}{R_1} \\ \\ \binom{R_1}{N} \end{matrix} \right\} \text{ — a distant view of the hill and the road,}$$

$\begin{pmatrix} N \\ E_2 \end{pmatrix}$

$\left. \begin{array}{c} \\ \\ \end{array} \right\}$ — my fingers adjust the binoculars, ·

$\begin{pmatrix} E_2 \\ IR \end{pmatrix}$

$\begin{pmatrix} IR \\ R_2 \end{pmatrix}$ — (almost) at the same time I am conscious of the movement of my fingers adjusting

$\begin{pmatrix} R_2 \\ N \end{pmatrix}$ the binoculars,

$\begin{pmatrix} N \\ E_1 \end{pmatrix}$ — I now see the road and the hill better, and I start on my way.

$\begin{pmatrix} E_1 \\ Obi \end{pmatrix}$

The matrix of couplings of the model is:

TABLE 5.6.1

Input	Output								Total of outputs
	from the outside	IR	R_1	R_2	N	E_1	E_2	Obi	
to the outside	·	1	1	1	1	1	1	1	7
IR	1	0	0	0	0	0	1	1	3
R_1	1	1	0	0	0	0	0	0	2
R_2	1	1	0	0	0	0	0	0	2
N	1	0	1	1	1	0	0	0	4
E_1	1	0	0	0	1	0	0	0	2
E_2	1	0	0	0	1	0	0	0	2
Obi	1	0	0	0	0	1	0	0	2
Total of inputs	7	3	2	2	4	2	2	2	24

5.7. Observation instruments coupled with a substitutive receptor

Instead of testing the temperature of water in a receptacle by plunging in my finger, which involves the risk of getting scalded, I prefer to look at a thermometer immersed in that water. A more complicated situation may also occur: I feel the water with my finger, but finding that method inadequate, I look at the thermometer immersed in the water and then manipulate the taps to adjust

FIGURE 5.7.0

Human being using an unregulated observation instrument coupled with a substitutive receptor

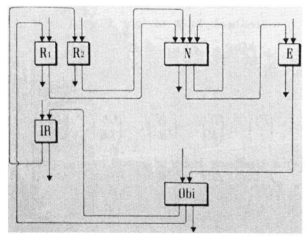

temperature to the desired level. The latter situation will be illustrated by a model (see Fig. 5.7.0). It is obvious that such model resembles in certain respects that discussed in Section 5.4 ("Performance instruments coupled with a substitutive effector").

The present model consists of six systems:

$$IR, R_1, R_2, N, E, Obi.$$

Attention is to be paid to the following feedbacks:

(1) the self-coupling of the central nervous system, which appeared in all the previous praxiological models and is mentioned each time to emphasize the memory function of reactions of the central nervous system,

$$\binom{N}{N};$$

(2) a feedback built of four throughputs:

$$\binom{R_1}{N}, \quad \binom{N}{E}, \quad \binom{E}{Obi}, \quad \binom{Obi}{R_1};$$

(3) a feedback built of five throughputs:

$$\binom{R_2}{N}, \quad \binom{N}{E}, \quad \binom{E}{Obi}, \quad \binom{Obi}{IR}, \quad \binom{IR}{R_2};$$

(4) a feedback built of seven throughputs:

$$\binom{R_2}{N}, \quad \binom{N}{E}, \quad \binom{E}{Obi}, \quad \binom{Obi}{R_1},$$

$$\binom{R_1}{N}, \quad \binom{N}{E}, \quad \binom{E}{Obi}, \quad \binom{Obi}{IR}, \quad \binom{IR}{R_2}.$$

It seems not out of place to give an example illustrating the feedbacks (2), (3) and (4). To the various systems of which the model consists the following rôles will be allotted:

(1) *Obi* — water in the receptacle and the taps,

(2) *IR* — the thermometer,

(3) R_1 — heat receptors of the hand,

(4) R_2 — the eyes (the set of receptors involved),

(5) N — the central nervous system,

(6) E — the hand adjusting the taps which admit cold and hot water (the set of effectors involved).

EXAMPLE I. The feedback (2) can also be symbolized thus (without changing the cyclic order):

$$\begin{pmatrix} Obi \\ R_1 \end{pmatrix}, \quad \begin{pmatrix} R_1 \\ N \end{pmatrix}, \quad \begin{pmatrix} N \\ E \end{pmatrix}, \quad \begin{pmatrix} E \\ Obi \end{pmatrix}.$$

It functions as follows:

$$\left.\begin{matrix} \begin{pmatrix} Obi \\ R_1 \end{pmatrix} \\ \begin{pmatrix} R_1 \\ N \end{pmatrix} \end{matrix}\right\}$$ — the hand tests the temperature of the water,

$$\left.\begin{matrix} \begin{pmatrix} N \\ E \end{pmatrix} \\ \begin{pmatrix} E \\ Obi \end{pmatrix} \end{matrix}\right\}$$ — the hand manipulates the taps.

EXAMPLE II. The feedback (3) can also be symbolized thus (without changing the cyclic order):

$$\begin{pmatrix} Obi \\ IR \end{pmatrix}, \quad \begin{pmatrix} IR \\ R_2 \end{pmatrix}, \quad \begin{pmatrix} R_2 \\ N \end{pmatrix}, \quad \begin{pmatrix} N \\ E \end{pmatrix}, \quad \begin{pmatrix} E \\ Obi \end{pmatrix}.$$

And it functions as follows:

$$\left.\begin{array}{c}\begin{pmatrix}Obi\\IR\end{pmatrix}\\[1em]\begin{pmatrix}IR\\R_2\end{pmatrix}\\[1em]\begin{pmatrix}R_2\\N\end{pmatrix}\end{array}\right\}$$ — the temperature of the water is checked (by means of the thermometer),

$$\left.\begin{array}{c}\begin{pmatrix}N\\E\end{pmatrix}\\[1em]\begin{pmatrix}E\\Obi\end{pmatrix}\end{array}\right\}$$ — the hand manipulates the taps.

EXAMPLE III. The feedback (4) can further be symbolized thus (without changing the cyclic order):

$$\begin{pmatrix}Obi\\R_1\end{pmatrix}, \begin{pmatrix}R_1\\N\end{pmatrix}, \begin{pmatrix}N\\E\end{pmatrix}, \begin{pmatrix}E\\Obi\end{pmatrix},$$

$$\begin{pmatrix}Obi\\IR\end{pmatrix}, \begin{pmatrix}IR\\R_2\end{pmatrix}, \begin{pmatrix}R_2\\N\end{pmatrix}, \begin{pmatrix}N\\E\end{pmatrix}, \begin{pmatrix}E\\Obi\end{pmatrix}.$$

It functions as follows:

$$\left.\begin{array}{c}\begin{pmatrix}Obi\\R_1\end{pmatrix}\\[1em]\begin{pmatrix}R_1\\N\end{pmatrix}\end{array}\right\}$$ — the hand tests the temperature of the water,

$$\left.\begin{array}{c}\begin{pmatrix}N\\E\end{pmatrix}\\[1em]\begin{pmatrix}E\\Obi\end{pmatrix}\end{array}\right\}$$ — the hand manipulates the taps,

$$\begin{pmatrix} Obi \\ IR \end{pmatrix}$$

$$\begin{pmatrix} IR \\ R_2 \end{pmatrix}$$ — the modified temperature of the water is checked by means of the thermometer,

$$\begin{pmatrix} R_2 \\ N \end{pmatrix}$$

$$\begin{pmatrix} N \\ E \end{pmatrix}$$ — the hand manipulates the taps again.

$$\begin{pmatrix} E \\ Obi \end{pmatrix}$$

EXAMPLE IV. There is nothing however to prevent us from symbolizing the feedback (4) in a different way, without changing the cyclic order of throughputs:

$$\begin{pmatrix} Obi \\ IR \end{pmatrix}'. \quad \begin{pmatrix} IR \\ R_2 \end{pmatrix}, \quad \begin{pmatrix} R_2 \\ N \end{pmatrix}, \quad \begin{pmatrix} N \\ E \end{pmatrix}, \quad \begin{pmatrix} E \\ Obi \end{pmatrix},$$

$$\begin{pmatrix} Obi \\ R_1 \end{pmatrix}, \quad \begin{pmatrix} R_1 \\ N \end{pmatrix}, \quad \begin{pmatrix} N \\ E \end{pmatrix}, \quad \begin{pmatrix} E \\ Obi \end{pmatrix}.$$

The description of the functioning will be modified correspondingly:

$$\begin{pmatrix} Obi \\ IR \end{pmatrix}$$

$$\begin{pmatrix} IR \\ R_2 \end{pmatrix}$$ — the temperature of the water is checked with the thermometer,

$$\begin{pmatrix} R_2 \\ N \end{pmatrix}$$

8

$\left.\begin{array}{c}\binom{N}{E}\\[2mm]\binom{E}{Obi}\end{array}\right\}$ — the hand manipulates the taps,

$\left.\begin{array}{c}\binom{Obi}{R_1}\\[2mm]\binom{R_1}{N}\end{array}\right\}$ — the hand tests the modified temperature of the water,

$\left.\begin{array}{c}\binom{N}{E}\\[2mm]\binom{E}{Obi}\end{array}\right\}$ — the hand manipulates the taps again.

The zero-one matrix of couplings of this model is shown in Table 5.7.1.

TABLE 5.7.1

Input	Output							Total of outputs
	from the outside	IR	R_1	R_2	N	E	Obi	
to the outside	.	1	1	1	1	1	1	5
IR	1	0	0	0	0	0	1	2
R_1	1	0	0	0	0	0	1	2
R_2	1	1	0	0	0	0	0	2
N	1	0	1	1	1	0	0	4
E	1	0	0	0	1	0	0	2
Obi	1	0	0	0	0	1	0	2
Total of inputs	6	2	2	2	3	2	3	20

5.8. A performance instrument plus an observation instrument

The analysis made so far (Sections 5.0 to 5.7) might — contrary to the author's intentions — lead the reader to conclude that for some incomprehensible reason we have tacitly assumed that the agent uses either a performance instrument or an observation instrument. But in fact the agent may for

FIGURE 5.8.0

Human being using both an unregulated observation instrument and an unregulated performance instrument

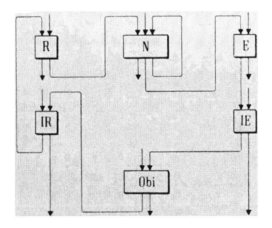

one and the same purpose use both a performance instrument and an observation instrument. To grasp this fully we may refer to a former example, mentioned when the concept of negative feedback was introduced (Section 2.8), namely that of a man

driving a car on a clear road and endeavouring
to keep it going at a constant speed. The driver
resorts to a performance instrument (the throttle,
which regulates the flow of petrol to the engine),
and an observation instrument (the speedometer).

To make quite sure that the reader may not
reach the erroneous conclusion mentioned above,
an outline is given of a certain praxiological
model (see Fig. 5.8.0) in which the agent uses both
an (unregulated) performance instrument and an
(also unregulated) observation instrument. Such
a model is obtained by coupling six elementary
systems:

$$IR, R, N, E, IE, Obi.$$

TABLE 5.8.1

Input	Output							Total of outputs
	from the outside	*IR*	*R*	*N*	*E*	*IE*	*Obi*	
to the outside	·	*1*	*1*	*1*	*1*	*1*	*1*	6
IR	*1*	*0*	*0*	*0*	*0*	*0*	*1*	2
R	*1*	*1*	*0*	*0*	*0*	*0*	*0*	2
N	*1*	*0*	*1*	*1*	*0*	*0*	*0*	3
E	*1*	*0*	*0*	*1*	*0*	*0*	*0*	2
IE	*1*	*0*	*0*	*0*	*1*	*0*	*0*	2
Obi	*1*	*0*	*0*	*0*	*0*	*1*	*0*	2
Total of inputs	6	2	2	3	2	2	2	19

Apart from the well-known self-coupling

$$\binom{N}{N}$$

we have to do here with the feedback:

$$\binom{IR}{R}, \quad \binom{R}{N}, \quad \binom{N}{E}, \quad \binom{E}{IE}, \quad \binom{IE}{Obi}, \quad \binom{Obi}{IR}.$$

The matrix of couplings is shown in Table 5.8.1.

5.9. Intellectual instruments

Our analysis of models in which an intellectual instrument (see Section 5.1) plays an essential rôle will also begin with examples.

EXAMPLE I. I have to make the choice between two possible routes for a journey I intend to make. The distance of *Route I* is D_1 miles and the price of the ticket is P_1 cents/mile, while the distance of *Route II* is D_2 miles and the price of the ticket is P_2 cents/mile. Before deciding, I have to compare the expense in the two cases, that is

(a) to calculate the product $D_1 \cdot P_1$,

(b) to calculate the product $D_2 \cdot P_2$,

(c) to compare the two products.

The operations (a) and (b) can be performed mentally, or alternatively with the aid of tables of products or a slide rule or an arithmometer.

EXAMPLE II. A designer is working on a project. His work requires certain calculations which, if simple enough, can be performed mentally, though

even so he may use tables, a slide rule or an arithmometer, merely to avoid fatigue or for greater
rapidity. In the case of all complicated calculations, the use of computation instruments becomes
imperative.

Every computation instrument, from the ancient
abacus to the modern automatic digital computer,
is, of course, an intellectual instrument. The reverse
statement does not hold, for while every dictionary
is an intellectual instrument, it is not a computation instrument.

A model of action involving the use of an intellectual instrument is given at Fig. 5.9.0.

FIGURE 5.9.0
Human being using an intellectual instrument

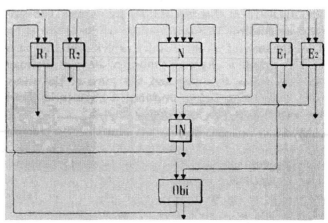

This is built of seven systems:

$$R_1, R_2, N, IN, E_1, E_2, Obi.$$

Attention must be drawn to the following feedbacks which occur in this model:

(1) the self-coupling

$$\binom{N}{N};$$

(2) the feedback built of four throughputs

$$\binom{R_1}{N}, \quad \binom{N}{E_1}, \quad \binom{E_1}{Obi}, \quad \binom{Obi}{R_1};$$

(3) the feedback built of four throughputs

$$\binom{R_2}{N}, \quad \binom{N}{E_2}, \quad \binom{E_2}{IN}, \quad \binom{IN}{R_2};$$

(4) the feedback built of five throughputs

$$\binom{R_1}{N}, \quad \binom{N}{N}, \quad \binom{N}{E_1}, \quad \binom{E_1}{Obi}, \quad \binom{Obi}{R_1};$$

(5) the feedback built of eight throughputs

$$\binom{R_1}{N}, \quad \binom{N}{E_2}, \quad \binom{E_2}{IN}, \quad \binom{IN}{R_2},$$

$$\binom{R_2}{N}, \quad \binom{N}{E_1}, \quad \binom{E_1}{Obi}, \quad \binom{Obi}{R_1}.$$

An example will elucidate these feedbacks. Let us consider a calculator who receives verbal instructions, makes .calculations with the aid of an arithmometer and verbally announces the result of his calculations. In this interpretation, the various systems involved in the model (see Fig. 5.9.0 and Table 5.9.1) will play the following rôles:

(1) R_1 — the calculator's ears (set of auditory receptors),

(2) R_2 — the calculator's eyes (set of visual receptors),

(3) N — the calculator's central nervous system,

(4) IN — the arithmometer,

(5) E_1 — the set of the calculator's vocal effectors,

(6) E_2 — the calculator's hands manipulating the arithmometer (the set of effectors involved),

(7) Obi — the person who gives the calculator verbal instructions and receives his answer (it may also be not a person, but an instrument, e. g., a magnetized tape with instructions for the machine and with a free space left for the result).

TABLE 5.9.1

Input	Output								Total of outputs
	from the outside	R_1	R_2	N	E_1	E_2	IN	Obi	
to the outside	.	1	1	1	1	1	1	1	7
R_1	1	0	0	0	0	0	0	1	2
R_2	1	0	0	0	0	0	1	0	2
N	1	1	1	1	0	0	0	0	4
E_1	1	0	0	1	0	0	0	0	2
E_2	1	0	0	1	0	0	0	0	2
IN	1	0	0	0	0	1	0	0	2
Obi	1	0	0	0	1	0	0	0	2
Total of inputs	7	2	2	4	2	2	2	2	23

The feedback (1) may function thus: The calculator has come across a possibility of simplifying
a certain kind of calculations and bears it in mind
for due application in practice.

The feedback (2) may function thus: The person
symbolized as "*Obi*" asks the calculator to solve
a certain computation problem, and the_ latter
replies.

The feedback (3): In solving his problem, the
calculator reads from the arithmometer a certain
intermediate result and inputs it into the arithmometer in order to obtain the final result.

The feedback (4): "*Obi*" gives the calculator
a task which requires several arithmetical operations, but is simple enough for the calculator to be
able to abstain from using the arithmometer. So
he solves the problem mentally and communicates
the result.

Finally, the feedback (5): "*Obi*" gives the
calculator a computational task, the latter
manipulates the arithmometer, reads the result
and communicates it verbally to "*Obi*".

5.10. Human co-operation

So far, all our praxiological models have imitated the behaviour of a single human agent, but
not that of a group of such agents. If the concepts
introduced in Chapters 1 and 2 are to be applicable to praxiology on a wider scale, it is essential

to build models imitating the behaviour of a group of agents. Exigencies of space require that we confine our analysis to a very simple model imitating the behaviour of two agents — an observer and a performer (see Fig. 5.10.0 and Table 5.10.2). The model of each agent constitutes a coupling of five elementary relatively isolated systems, viz., two receptors, one central nervous system, and two effectors (for details see Table 5.10.1).

FIGURE 5.10.0

Human co-operation: an observer and a performer

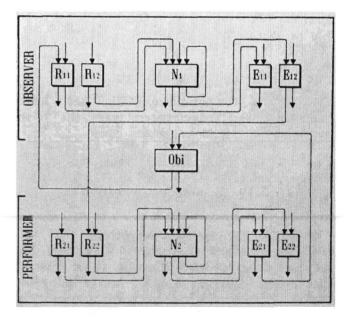

TABLE 5.10.1

Type of elementary system	Observer (Person no. I)	Performer (Person no. II)
Receptor used to observe the object of action (e. g., equivalent of the eye)	R_{11}	R_{21}
Receptor used to receive information from the other person (e. g., equivalent of the ear)	R_{12}	R_{22}
Central nervous system	N_1	N_2
Effector used to influence the object of action (e. g., equivalent of the hand)	E_{11}	E_{21}
Effector used to give information to the other person (e. g., equivalent of the vocal organ)	E_{12}	E_{22}

Attention is directed to the following feedbacks in this model:

(1) the self-coupling which consists of the throughput

$$\binom{N_1}{N_1},$$

(2) the self-coupling which consists of the throughput

$$\binom{N_2}{N_2},$$

(3) the "big loop" of the model, built of the throughputs

$$\binom{R_{11}}{N_1}, \quad \binom{N_1}{E_{12}}, \quad \binom{E_{12}}{R_{22}},$$

$$\binom{R_{22}}{N_2}, \quad \binom{N_2}{E_{21}}, \quad \binom{E_{21}}{Obi}, \quad \binom{Obi}{R_{11}}.$$

TABLE 5.10.2

Input	from the outside	Output											Total of outputs
		R_{11}	R_{12}	N_1	E_{11}	E_{12}	Obi	R_{21}	R_{22}	N_2	E_{21}	E_{22}	
to the outside	.	1	1	1	1	1	1	1	1	1	1	1	11
R_{11}	1	0	0	0	0	0	1	0	0	0	0	0	2
R_{12}	1	0	0	0	0	0	0	0	0	0	0	0	1
N_1	1	1	1	1	0	0	0	0	0	0	0	0	4
E_{11}	1	0	0	1	0	0	0	0	0	0	0	0	2
E_{12}	1	0	0	1	0	0	0	0	0	0	0	0	2
Obi	1	0	0	0	0	0	0	0	0	0	1	0	2
R_{21}	1	0	0	0	0	0	0	0	0	0	0	0	1
R_{22}	1	0	0	0	0	1	0	0	0	0	0	0	2
N_2	1	0	0	0	0	0	0	1	1	1	0	0	4
E_{21}	1	0	0	0	0	0	0	0	0	1	0	0	2
E_{22}	1	0	0	0	0	0	0	0	0	1	0	0	2
Total of inputs	11	2	2	4	1	2	2	2	2	4	2	1	35

It does not seem necessary to discuss the functions of such feedbacks, since the reader himself will now easily find suitable interpretations.

5.11. Automata

The colloquial usage of the term "automaton" is somewhat vague. A precise definition seems advisable, since mankind has entered the epoch in which intellectual instruments (see Section 5.9) and automata are beginning to play a vast rôle in industrial production and scientific research. Let it also be borne in mind that most of the latest intellectual instruments are automata (e. g., automatic electronic digital computers).

No standard definition of automaton, such as would meet all the requirements of modern methodology, will be introduced here. Exigencies of space demand that we be satisfied with a more modest procedure.

Many readers may say: "Fair enough for a section on the concept of automaton to be included in a book on cybernetics, but why in a chapter on praxiological models?" To answer this, a somewhat circuitous route will be followed: we shall first establish what should be the structure of models of automata, and then try to indicate — what are automata.

The simplest praxiological model (Section 5.1, Fig. 5.1.0) is built of only four systems:
(1) R — receptor,
(2) N — central nervous system,
(3) E — effector,
(4) Obi — object of action.

In such a model, the system obtained by the serial coupling of the systems R, N and E represents an individual man, and the system Obi, his environment or a part of that environment. Thus a system built of the systems R, N and E by means of the throughputs

$$\binom{R}{N}, \quad \binom{N}{E}, \quad \cdot\cdot$$

to be called briefly *"RNE"*, is a most radically simplified model of any man and, moreover, of many an animal (Fig. 5.1.0). But *RNE* can also play another rôle — that of a most radically simplified model of any automaton. In such a case the rôles of the various systems involved in the model (Fig. 5.1.0, as before) change considerably:

(1) R — an information system (e. g., a photo-electric cell, a microphone) which receives information about the system *Obi* and transmits it to the system N;

(2) N — an information system which receives information from the system R, transforms it into information-decisions and sends them to the system E;

(3) E — an informed system which receives information-decisions from the system N and acts on the system *Obi* in conformity with such decisions;

(4) *Obi* — an informing system which is outside the automaton *RNE* and is "under the observation" of the system R and "under the pressure" of the system E.

It will be readily appreciated that a praxiological model (Fig. 5.1.0, as before) interpreted in this way can serve as a radically simplified model of a slot machine (selling tickets, sweets, etc.) and of an automaton which controls a machine tool.

Now the concept of automaton will be characterized by two postulates:

(1) Every automaton is built exclusively of inanimate technical matter (type D in Section 3.1).

(2) Every automaton is obtained by a serial coupling of prospective systems of at least three kinds:

(a) information systems, i. e., pseudo-receptors obtaining information from the outside;

(b) information systems, i. e., pseudo-central--nervous-systems, obtaining information from pseudo--receptors and transforming it into information-instructions;

(c) informed systems, i. e., pseudo-effectors obtaining instructions from the inside of the automaton and correspondingly acting on the outside of the automaton.

It is not our intention to ascribe any particular importance to these postulates; but it just seems better to be guided by them than by an intuitive concept taken from the colloquial usage.

Several points here warrant a brief mention:

(1) The meaning of the term "automaton", as defined by Postulates (1) and (2), is broad enough to cover:

(a) automata endowed with conditioned responses (Section 4.1);

(b) automata so coupled with one another that one "teaches" the other, or they teach each other reciprocally (Section 4.3).

(2) In conformity with Postulates (1) and (2), each synthetic animal (Section 4.0) can be considered an automaton.

(3) In view of Postulate (1) it does not seem possible to regard as an automaton any plant, any animal, any human being, or any animate technical matter (Section 3.1).

(4) Postulates (1) and (2) certainly do not suffice to answer the question as to whether all or some automata think. For the time being, we know too little not only to be able to answer that question, but even to understand it in an unambiguous way. The next chapter will perhaps somewhat improve our position in that difficult matter.

(Cf. *Ref.* B. 1, C. 1, L. 3)

5.12. Concluding remarks

The object of this chapter was twofold: to show that the basic cybernetic concepts are useful in praxiology, and to shed at least some light on the relationship between cybernetics and praxiology, i. e., to study the connections between the ideas of Norbert Wiener (cybernetics) and those of Tadeusz Kotarbiński (praxiology) (cf. *Ref.* K. 1).

Neither object is particularly easy to achieve, especially in a popular form. A systematic exposition has been replaced by a number of examples, concerning which the reader may object that, even if taken together, they do not form an exhaustive whole, that they are excessively simplified, if not trivial, and that too much abstraction marked the models of the given concrete examples.

But, on the other hand, a careful reader will be able to build for himself more complicated and correspondingly more interesting models, for instance such in which there are more complex connections between the agents than those described in Section 5.10; or such in which apart from a group of agents there are not only such instruments as establish couplings between the agent and the object of action, but also such as establish couplings between the different agents (a communication network). Perhaps the analysis of economic models (presented later on) will be instructive to those readers who are interested chiefly in praxiological models.

One remark more: for lack of space we have been unable to discuss the important problem of how praxiological models, built in conformity with the basic cybernetic concepts, may serve as foundations for a general theory of prostheses (cf. *Ref.* GBS and W. 2).

6. SIGNALS AND EXPRESSIONS, CODES AND LANGUAGES

6.0. Introductory remarks

When discussing, in the preceding chapter, some of the praxiological models, we referred to the system of which individual models were built, and concentrated on the inputs and outputs (in particular throughputs) of such models. We disregarded, however, calendars and repertories, trajectories and determinators (cf. Sections 1.1 and 1.9). To fill at least some of those gaps we shall now deal with the *repertories* of certain inputs and outputs, namely some information inputs and outputs (Section 2.0) of praxiological models.

In building praxiological models we had to do with, among other things, receptors, central nervous systems, effectors, observation instruments (Sections 5.5, 5.6, 5.7, 5.8) and intellectual instruments (Section 5.9). It is readily appreciated that:

(1) every receptor is an information system,

(2) every central nervous system is an information system,

(3) every effector is an informed system,

(4) every observation instrument is an informing system, and

(5) every intellectual instrument is an information system.

It is obvious that each of the statements (1) to (3) is satisfied not only by any (true) receptor, central nervous system or effector built of animate non-technical matter, but also by every model built of inanimate technical matter, or even animate technical matter (Sections 3.1 and 3.7).

Consequently, the distinguishable states of many inputs and outputs of the praxiological models are simply *information*. But not every information occurring in the systems in question is simply a distinguishable state of an input or an output, since there exist much more complex forms of information, information being a "parcel" of distinguishable states. To discuss their structure would be outside the scope of this book.

6.1. |Ordered pairs

The concept of ordered pair probably owes its emergence to analytical plane geometry (since every point on a Euclidean plane is univocally determined by an ordered pair of its co-ordinates). It is a concept which has proved useful in theoretical arithmetic (complex numbers as ordered pairs of real numbers, fractions as ordered pairs of integers), and from theoretical arithmetic it found its way into the set theory and logic, where it has acquired necessary precision and generality (cf. *Ref.* KM p. 51, K. 2, M. 3 pp. 148 and 359, W. 1).

The basic property of the concept of ordered pair is as follows: The ordered pair consisting of the elements x_1, y_1 is identical with the ordered pair consisting of the elements x_2, y_2 if, and only if, the following two conditions are satisfied: (1) x_1 is identical with y_1, and (2) x_2 is identical with y_2.

Thus the ordered pair consisting of Napoleon I and Napoleon III is identical with the ordered pair consisting of the victor of Jena and the loser of Sedan. But the ordered pair consisting of Napoleon I and Napoleon III is *not* identical with the ordered pair consisting of the loser of Sedan and the victor of Jena, because

(1) Napoleon I is not identical with the loser of Sedan,

(2) Napoleon III is not identical with the victor of Jena.

Moreover, if we have two items, for instance two different persons, John and Peter, we can build as many as *four* different ordered pairs, viz.:

(1) John, John;
(2) John, Peter;
(3) Peter, John;
(4) Peter, Peter.

Of course, such consequences of what is universally accepted in mathematics are somewhat awkward from the materialistic point of view. If only matter and bodies, conceived as fragments of matter, exist then it is difficult to believe that there are two *different* bodies one of which consists of John and Peter, and the other of Peter and John.

Such difficulties can, nevertheless, be surmounted to a considerable degree, but we shall not concern ourselves here with a generalized materialistic interpretation of the term "ordered pair", since it would here be superfluous. We confine ourselves to the interpretation of a very special situation:

(1) two kinds of objects, kind I and kind II, are given, these two kinds being disjoint, which means that no object of kind I is an object of kind II;

(2) no object of kind I is part of any object of kind II;

(3) we take into consideration only ordered pairs in which the first element belongs to kind I and the second to kind II.

Given such assumptions, one ordered pair, and only one, can be built of two objects when one of such is of kind I and the other of kind II. Consequently, in a situation characterized by the conditions (1), (2) and (3), an ordered pair can be treated in a purely materialistic way as a combination of two disjoint bodies, one of which is of kind I, and the other of kind II.

Now, e. g., if the objects of kind I are exclusively human beings, and the objects of kind II are exclusively dogs, and if John and Peter are two different persons, and Brisky and Frisky two different dogs, then:

the ordered pair consisting of John and Brisky is not identical with an ordered pair consisting of John and Frisky, since Brisky is not identical with Frisky;

the ordered pair consisting of John and Brisky is not identical with an ordered pair consisting of

Peter and Brisky, since John is not identical with Peter;

a fortiori, the ordered pair consisting of John and Brisky is not identical with an ordered pair consisting of Peter and Frisky.

Apart from the concept of ordered pair, other "parcel-like" concepts are used in modern logic and mathematics, such as the concept of ordered triple, ordered quadruple, ordered *n*-tuple, matrix, etc. We can also build "parcels" consisting of "parcels". All this, however, must be passed over here for lack of space.

6.2. Signals and codes

We shall refer briefly to *"elementary signal"* instead of saying "information which is a distinguishable state", and we shall speak of *"compound signal"* instead of saying "information which is a parcel of distinguishable states".

It seems advisable to give simple examples: if we take into consideration a zero-one information system (cf. 2.2 and 2.5), e. g., an alternative system, and number its inputs (input I and input II), then the distinguishable state *0* and the distinguishable state *1* are elementary signals, and the ordered pair built — e. g., of the element *0* (the state of input I) and the element *1* (the state of input II) — is a compound signal.

Certain systems of signals will be called *"codes"*. Each code is either the repertory of the distinguishable states of the given information input or output, or the set of all compound signals that can, in conformity with given rules, be built of a given repertory — or repertories — of elementary signals. In pracitce, an important rôle is played by binary codes, i. e., codes built of two elementary signals.

6.3. Expressions and languages

Some, but not all, signals are expressions, but all expressions are signals.

Some, but not all, codes are languages, but all languages are codes.

By *"language"* we mean such code, and only such, which consists entirely of expressions.

Of course, the definition given above is useful only if we are in a position to distinguish expressions from signals that are not expressions. Three kinds of situations are involved here:

(1) easily ascertainable — that a given signal *is* an expression;

(2) easily ascertainable — that a given signal *is not* an expression;

(3) dubious cases.

Ad (1). Peter *says* to Paul: "Our bus is coming". What Peter has said is, of course, an expression. Peter *writes* to Paul: "I had fluflu'". What Peter has written is, of course, an expression. A policeman *indicates* to pedestrians that they may cross the street. His gesture too is considered to be an expression.

Ad (2). Signals are incessantly passing from Peter's receptors to his central nervous system, and his central nervous system is incessantly sending signals to Peter's effectors. *None* of these signals is an expression.

Ad (3). If instead of a policeman regulating the traffic we have automata with lights in three colours, then whichever of those lights is actually lit (green, amber, or red) is undoubtedly a signal belonging to a certain code. But is it also an expression belonging to a certain language? Here, intuition is not enough. The thermometer shows the temperature of its surroundings, the watch indicates time. Both the thermometer and the watch give signals belonging to certain codes (the code of the thermometer scale and the code of the watch dial, respectively). That much is quite obvious. But consider the signals at the throughputs from observation instruments to receptors (Sections 5.5, 5.6, 5.7, 5.8); are all or some of them expressions? An analogous question arises as regards the outputs and inputs of intellectual instruments (Section 5.9). Moreover, in certain cases, an analogous question might be asked with respect to the control inputs of performance instruments, i. e., such inputs as the motorcar throttle or the gear lever (Section 5.3).

Let us suppose that we have taken two decisions. First, each light *signal* at the crossing is considered an *expression* of a certain language. Second, each movement of the driver's foot which changes the position of the throttle and each movement of his hand which changes the position of the gear lever is considered not only a *signal* given to the machine and belonging to a certain code, but also an *expression* belonging to a certain language. Suppose further, that these decisions of ours have only terminological value. And now let us look into the future. Drivers have been eliminated, their eyes have been replaced by systems of photocells, their central nervous systems have been replaced by computers the perfomance organs of which have replaced driver's hands and feet. Are we, in view of the previous terminological decisions, obliged, or perhaps only inclined, to consider such light signals, passing from the traffic-regulating automaton to the photoelectric cells of the car, as expressions? The same refers to the movements performed by the devices which constitute the outputs of the automaton.

These questions are left unanswered. Perhaps the reader will himself attempt certain solutions when he is more familiar with the internal structure of certain signals, namely semantic expressions.

(Cf. *Ref.* B. 1, B. 2, S. 1, S. 2)

6.4. Manifested semantic expressions as ordered pairs

Peter *says*: "Warsaw lies in Europe, and Peking lies in Asia". This information, spoken by Peter, is an expression which is in turn built of other expressions, such as the expression which *sounds* "Warsaw", the expression which sounds "lies", etc.

Peter *writes*: "Warsaw lies in Europe, and Peking lies in Asia". This information, written by Peter, is an *expression* which in turn is built of other expressions, such as the *written* expression "Warsaw", the written expression "lies", etc.

Peter *draws* a map of Eurasia, that is, he draws the contour of the Eurasiatic continent, draws the line which demarcates Europe from Asia, and marks the positions of Warsaw and Peking. That map, drawn by Peter, also is an expression, built of other expressions — produced by *drawing* — such as the contour, the demarcation line, the two dots.

Peter *grimaces* because his tooth aches. The position and/or *movements* of his facial muscles, or of any other man's, will also be called an expression, although in this case we abstain from discussing whether such an expression, too, is built of other expressions.

Peter *conducts* an orchestra, his hands make certain movements. Every *gesture* by any man will also be called an expression.

Every expression described above is a manifested semantic expression. Every manifested semantic expression is an ordered pair, the first element of which we shall call its "*transmission element*", and the second, its "*semantic element*".

The transmission element of any manifested semantic expression is always a distinguishable state (or a "parcel" of distinguishable states) of the ouputs of some effectors or of the outputs of some performance instruments. The transmission element of an expression spoken by a man is a certain sequence of sound waves (the surrounding air being in that case the performance instrument). The transmission element of an expression written is a certain complicated solid built of bits of chalk, dried ink, dried paint, etc. The transmission element of a gesture is a certain sequence of states of certain muscle effectors.

The semantic element always consists in a certain distinguishable state of a certain input or output of a human central nervous system. This does not imply that, *vice versa*, every distinguishable state of at least some inputs and outputs of a human central nervous system is a semantic element of some expression.

6.5. Unmanifested semantic expressions as ordered pairs

If a man talks to himself — thinks, that is — he does by way of certain imagined expressions, i. e., he conjures up certain inscriptions (e. g., mathematical or chemical formulae), certain graphs, maps, etc. The semantic elements of such expressions involve no difficulty or doubts: they correspond to those of expressions actually written or drawn. But what in such cases are the transmission elements? We assume that the transmission element of every imagined expression is a certain state of a human central nervous system; and at the same time we assume that *no* such state of any human central nervous system which is a transmission element of an unmanifested (imagined) semantic expression is a semantic element of *any* expression. Consequently, in every expression we can easily distinguish the first — transmission — element, from the second — semantic — element. This assumption enables us so to extend our conception of treating manifested semantic expressions as ordered pairs as to cover also all the unmanifested semantic expressions.

6.6. Semantic thinking and semantic communication

For one who is not a psychologist, it is practically impossible to describe the processes of thinking and of interhuman communication. But we shall nevertheless endeavour to present some of these processes in a radically simplified way, by means of a model, similar to those described in Chapter 5, and in particular to the praxiological model of interhuman co-operation (Section 5.12). If that model of semantic thinking and semantic communication proves to be faulty from the psychological point of view, it will at least have the merit of giving rise to sound professional criticism.

This model (see Fig. 6.6.0) consists of seven systems:

(1) R_1 — receptor
(2) N_1 — central nervous system } Person I
(3) E_1 — effector

(4) R_2 — receptor
(5) N_2 — central nervous system $\Big\}$ Person II
(6) E_2 — effector
(7) B — a blackboard (properly placed and lighted, with a piece of chalk and a sponge).

The model functions as follows:

(1a) *Person I* thinks, without manifesting what he thinks. Here, the process of thinking is a se-

FIGURE 6.6.0
Reciprocal communication

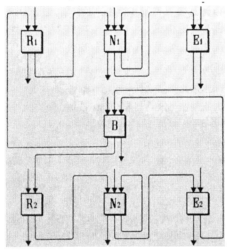

quence of unmanifested semantic expressions. Each such expression is an ordered pair, consisting of the transmission element and the semantic element, which is the distinguishable state of the throughput

$$\begin{pmatrix} N_1 \\ N_1 \end{pmatrix}.$$

(1b) *Person II* behaves in the same way as does Person I. The analogy is complete but for the fact that this time we have to do with the throughput

$$\begin{pmatrix} N_2 \\ N_2 \end{pmatrix}.$$

(2a) *Person I* thinks and fully manifests what he thinks. His process of thinking is a sequence of manifested semantic expressions. Each such expression is an ordered pair consisting of the transmission element (e. g., an inscription on the blackboard B performed by means of the effector E_1) and the semantic element which is the distinguishable state of the throughputs

$$\text{at first } \begin{pmatrix} N_1 \\ E_1 \end{pmatrix}, \quad \text{and then } \begin{pmatrix} R_1 \\ N_1 \end{pmatrix}.$$

The following feedback occurs here:

$$\begin{pmatrix} N_1 \\ E_1 \end{pmatrix}, \quad \begin{pmatrix} E_1 \\ B \end{pmatrix}, \quad \begin{pmatrix} B \\ R_1 \end{pmatrix}, \quad \begin{pmatrix} R_1 \\ N_1 \end{pmatrix}.$$

(2b) *Person II* behaves in the same way as does *Person I*. The analogy is complete but for the fact that this time we have to do with the feedback:

$$\begin{pmatrix} N_2 \\ E_2 \end{pmatrix}, \quad \begin{pmatrix} E_2 \\ B \end{pmatrix}, \quad \begin{pmatrix} B \\ R_2 \end{pmatrix}, \quad \begin{pmatrix} R_2 \\ N_2 \end{pmatrix}.$$

(3a) *Person I* conveys some information to *Person II*. We have here to do with a sequence of manifested semantic expressions issued by *Person I* and a sequence of manifested semantic expressions received by *Person II*. Any expression issued by

Person I is an ordered pair consisting of the transmission element and the semantic element. The latter is the distinguishable state of the throughput

$$\begin{pmatrix} N_1 \\ E \end{pmatrix},$$

while the former is the distinguishable state of the throughput

$$\begin{pmatrix} B \\ R_2 \end{pmatrix}.$$

If we take into account the expression received by *Person II*, that expression also consists of the transmission element and the semantic element. The former is the same as before, i. e., the distinguishable state of the throughput

$$\begin{pmatrix} B \\ R_2 \end{pmatrix},$$

while the latter is a certain distinguishable state of the throughput

$$\begin{pmatrix} R_2 \\ N_2 \end{pmatrix}.$$

Thus, we deal here with the transmission element which occurs once in the same throughput, and the semantic element which is common to both persons (provided there has been no misunderstanding between them), but occurs twice, each time in a different throughput.

(3b) *Person II* conveys some information to *Person I*. The analogy is complete but for the fact that this time the semantic element occurs in the throughput

$$\begin{pmatrix} N_2 \\ E_2 \end{pmatrix},$$

the transmission element occurs in the throughput

$$\begin{pmatrix} B \\ R_1 \end{pmatrix},$$

and the semantic element occurs again in the throughput

$$\begin{pmatrix} R_1 \\ N_1 \end{pmatrix}.$$

(4) When *Person I* and *Person II* exchange information, we observe the following feedback:

$$\begin{pmatrix} N_1 \\ E_1 \end{pmatrix}, \quad \begin{pmatrix} E_1 \\ B \end{pmatrix}, \quad \begin{pmatrix} B \\ R_2 \end{pmatrix}, \quad \begin{pmatrix} R_2 \\ N_2 \end{pmatrix},$$

$$\begin{pmatrix} N_2 \\ E_2 \end{pmatrix}, \quad \begin{pmatrix} E_2 \\ B \end{pmatrix}, \quad \begin{pmatrix} B \\ R_1 \end{pmatrix}, \quad \begin{pmatrix} R_1 \\ N_1 \end{pmatrix}.$$

6.7. Formal expressions

Peter, who is British, has a friend Jean, who is French and who does not understand a word of English. In his presence Peter says "John is a fathead", and Jean repeats carefully after him, "John is a fathead", without understanding what he says. In this case we have two expressions which sound identically, yet what Peter has spoken is a semantic expression, and what Jean has spoken is not. (Should another English-speaking person, e. g., George, hear what Jean has said, then we would have to do with a semantic expression, but a received semantic expression, and not an issued semantic expression. The semantic element would be found in George's central nervous system, but would be absent from Jean's central nervous system.) Peter has uttered a (manifested) semantic expression, and Jean has uttered a formal expression. If any one says or writes "*Woggles diggle*" or "*Woggled diggles miggle*", he produces a formal expression. Apart from semantic and formal expressions, there are also mixed semantic-formal expressions.

This concludes our explanations or rather a perhaps clumsy endeavour to give a precise meaning to the concept of expression. We have certainly not endowed the term "expression" with any sharply-defined meaning, since there are cases in which our explanations do not provide a sufficient basis for deciding whether we have to do with an expression or with an object which is not an expression. For instance, we are not in a position to decide whether the inarticulate sounds produced by an insane person or the stammering of a drunkard consists, at least in part, of expressions. Yet our explanations are sufficient to indicate, in many cases, that something is, or is not, an expression, and even such a modest result will prove useful in further analysis.

10

6.8. Formal operations performed on expressions

Much of our intellectual work; such as writing or editing a text, translation from one language into another, computations, etc., consists in performing on certain expressions such operations as will always give new expressions. Among those operations which lead from expressions to expressions we find a certain kind of operations — formal operations — which are for us of particular interest.

We say briefly that a given operation is a formal operation instead of saying that it satisfies all the following conditions:

(1) it is performed exclusively on expressions,

(2) its result is always an expression,

(3) in order to perform it on semantic expressions, it is not necessary to know the semantic element of any expression involved.

(4) in order to perform it on some expressions, whether semantic or formal, it is yet necessary to know at least some syntactic properties of all the expressions involved.

Formal operations are remarkable for the important property that they can be carried out by a machine, i. e., a technical relatively isolated system. (Obviously, the construction of such a machine is not always technically possible, and if technically possible, is not always justified on economic grounds.) A machine for performing formal operation(s) must be built so that when we input the expressions on which the operation(s) is (are) to be performed, the expression which is the result of the operation(s) will appear at the output. For example, in the case of an arithmometer we input the items to be summed, and the sum appears at the output

If more examples are desired, we can specify at least five spheres of intellectual activity, some of them apparently quite remote from one another, in which formal operations appear:

(1) confrontation of contradictory information,

(2) inference, especially deductive inference,

(3) numerical computations,

(4) translation from one language into another,

(5) artistic compositions.

(1)-(4) will be dealt with in greater detail in the two following chapters.

Attention must also be drawn here to two operations, viz., formalization and semantic interpretation.

Formalization consists in depriving a semantic expression of its semantic element. The result is always a formal expression, and formalization itself is a formal operation.

Semantic interpretation consists in imparting to a formal expression a semantic element, that is, in transforming a formal expression into a semantic one. (Cf. *Ref.* G. 4)

6.9. Semantic languages, formal languages, and mixed semantic-formal languages

A semantic language is every language, and only such, which consists exclusively of semantic expressions (Sections 6.4 and 6.5).

A formal language is every language, and only such, which consists exclusively of formal expressions (Section 6.7).

A semantic-formal language is every language, and only such, which consists of both semantic and formal expressions, and possibly also of mixed semantic-formal expressions.

It might seem at first glance that every natural language (i. e., a language of a nation or a tribe) is a semantic language. It might also seem at first glance (to those not familiar with modern logic) that the concept of formal language is a strange and useless one.

In fact, it turns out that formal expressions occur in natural languages, so that at least some natural languages are not purely semantic, but mixed semantic-formal languages. As examples may be quoted meaningless expressions often occurring in conversation, such as "Well you know", and certain made-up words.

The concepts of formal expression and formal language have emerged in connection with the modern logical research on the foundations of mathematics. They have also proved useful in cybernetics, that is in the modelling of certain intellectual activities and in the building of automata. These matters will be dealt with in the next chapter. Here, an example of a formal language will be given.

Suppose there are three states: Aland, Beland and Celand. Aland has her embassies in Beland and Celand, and her Ministry of Foreign Affairs exchanges with them messages in cipher. For some reason, it has become necessary to change the cipher, and the Ministry orders such from an appropriate outstanding expert. But the Ministry does not want that expert to be in a position in the future to decipher the messages; hence it asks him to work out a formal language. He is to conceive a store of words of the new formal language, together with the grammar of that language, emphasis being laid on syntax. The store of words must be rich enough, both quantitatively and qualitatively (parts of speech), to make it possible to conduct official correspondence in the new language. When the expert has done his job and submitted the dictionary and the grammar of the new language, the Ministry, without informing the expert, gives a semantic interpretation to the formal language (Section 6.8). That language is then turned into a semantic or a mixed semantic-formal language. But there is nothing to prevent the Ministry working out several different semantic interpretations, to be used on different occasions, e. g.:

the first for its messages to the embassy in Beland,

the second for messages from the embassy in Beland to the Ministry,

the third for its messages to the embassy in Celand,

the fourth for messages from the embassy in Celand to the Ministry.

Further semantic interpretations can be worked out in the course of time.

6.10. True expressions, singled out expressions, necessarily true expressions

In any semantic language there may be distinguished sentential expressions (some or all of which can be called "sentences"), among which may in turn be distinguished true expressions, i. e., expressions which describe facts in conformity with facts. The two following expressions belong to a certain semantic language (viz., English):

(1) Warsaw is not the capital of Poland,
(2) Warsaw is the capital of Poland.

They are sentential expressions (viz., sentences), but only expression (2) is true.

By analogy to the distinction between true and other expressions in semantic languages, there may, in the case of formal languages, be introduced, a distinction into *singled out* and other expressions. We can imagine a language in which every expression is built of five letters, each letter being either "*A*" or "*B*". Thus, among its expressions we shall have such as

$$AAAAA, \; BBBBB, \; ABBBA, \ldots$$

We shall single out all those expressions in which all the odd letters are "*B*", e. g.,

$$BBBBB, \; BABAB, \; BBBAB, \ldots$$

In the mixed semantic-formal languages an analogous distinction is more complicated, since three categories of expressions are there distinguished:

(1) true expressions,

(2) singled out expressions,

(3) necessarily true expressions.

Ad (3). By a necessarily true expression we mean every expression, and only such, of a given mixed semantic-formal language which satisfies the following two conditions:

(1) it is built of at least one semantic expression and at least one formal expression (Section 6.9),

(2) for any semantic interpretation (the syntax of the language remaining the same) that expression becomes true.

An example: Take into account the language L, which is the English language enriched with two formal expressions:

a pseudo-noun   *woggle*,

a pseudo-intransitive verb *diggle*.

The syntax of L is the syntax of the English language. The following expression, which belongs to L, is a necessarily true expression:

If *woggle diggles* then *woggle diggles*.

6.11. Formalization, modelling, automatization

It is not intended to enlarge here on the importance of formalization in research on the foundations of mathematics (Section 6.8) and on the limits of formalization, first pointed out by the Austrian logician Kurt Gödel; it will be sufficient to confine ourselves to elementary problems of modelling (Section 3.7).

Suppose that we are required to model a certain intellectual activity carried out by a certain individual. That operation is usually performed in some semantic language (Section 6.9). Modelling is no easy matter here, since the model has to cover both the transmission elements and the semantic elements of the semantic expressions involved (Sections 6.4 and 6.5). True, we have already had to do with such a modelling (Section 6.6), but the modelled processes were very simple, and the model itself was outlined very superficially.

Yet, in our opinion, in cybernetic research we should not renounce modelling intellectual activities performed in semantic languages; moreover, the modelling of such activities, and perhaps even of all psychic processes, is an extremely important aspect of such research (although it is somewhat difficult to say to what extent it can be carried out). But if the intellectual process to be modelled can be reduced to performance of formal operations (Section 6.8), if the semantic language in which the given process is carried out can be replaced by a formal language, and if true expressions can be replaced by singled out expressions (Section 6.10), then the task of modelling such a process is radically simplified (e. g., in the case of numerical calculus), and the model becomes so easy from the technological point of view as to be a prototype of an automaton (e. g., a programmed digital computer). It is precisely in this point that formalization is connected with automatization.

7. L O G I C A L   M O D E L S

7.0. Introductory remarks

This chapter deals with the modelling of certain intellectual functions (processes) which are reducible to formal operations (Section 6.8), such as:

(1) confrontation of contradictory information,
(2) arithmetical operations,
(3) translation from one language into another.

7.1. Confrontation of contradictory information

In describing the praxiological models (in Chapter 5), attention was paid to the unreliability of receptors and observation instruments. Hence, every receptor and every observation instrument was always provided with an additional input. There exist well-known methods of reducing information errors originating in the additional inputs to receptors and observation instruments. Such methods are, as a rule, statistical and probabilistic; although not fully reliable, they are of considerable practical importance. Without describing them in detail, it may here be stated that they are always based on at least one of the following two procedures:

collection of simultaneous data from different receptors or observation instruments,

collection of consecutive data from the same receptor or observation instrument.

Further work on a set of simultaneous or consecutive items of information, each of which may include some error, depends on the character of that information. If we are dealing with, for example, a set of independently made measurements, then the most reasonable thing to do is to take, instead of any individual measurement, the arithmetical average or the median. If we have to do with a set of independent items of qualitative information, each item being either "yes" or "no", then in certain cases the best thing is to follow the majority (of affirmative or negative items of information).

The modelling of such operations does not involve any major difficulties. We shall outline here only one model (Fig. 7.1.0), which is built of the following seven systems:

(1) R_1 } three receptors supplying simultaneous,
R_2 — } independent items of information, not
R_3 — } necessarily agreeing with one another;

(4) C — organ of the central nervous system which confronts information supplied by R_1, R_2, R_3;

(5) D — organ of the central nervous system which takes decisions in conformity with the result of confrontation;

(6) E — effector;

(7) Obi — object of action.

FIGURE 7.1.0
Confrontation of data before a decision is made

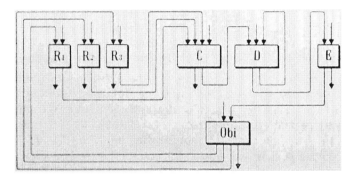

Since the model is very simple, we abstain from considering its functioning in detail, leaving that to the reader. This refers in particular to the functioning of the three analogous feedbacks:

(1) $\quad \begin{pmatrix} R_1 \\ C \end{pmatrix}, \begin{pmatrix} C \\ D \end{pmatrix}, \begin{pmatrix} D \\ E \end{pmatrix}, \begin{pmatrix} E \\ Obi \end{pmatrix}, \begin{pmatrix} Obi \\ R_1 \end{pmatrix};$

(2) $\quad \begin{pmatrix} R_2 \\ C \end{pmatrix}, \begin{pmatrix} C \\ D \end{pmatrix}, \begin{pmatrix} D \\ E \end{pmatrix}, \begin{pmatrix} E \\ Obi \end{pmatrix}, \begin{pmatrix} Obi \\ R_2 \end{pmatrix};$

(3) $\quad \begin{pmatrix} R_3 \\ C \end{pmatrix}, \begin{pmatrix} C \\ D \end{pmatrix}, \begin{pmatrix} D \\ E \end{pmatrix}, \begin{pmatrix} E \\ Obi \end{pmatrix}, \begin{pmatrix} Obi \\ R_3 \end{pmatrix}.$

7.2. Parallely confronting system

We now make certain additional assumptions concerning the model investigated in the preceding section (Fig. 7.1.0). First of all, we assume that each of the three receptors, R_1, R_2, R_3, supplies exclusively qualitative information, either affirmative or negative. This means simply that in our model each of the throughputs

$$\binom{R_1}{C}, \quad \binom{R_2}{C}, \quad \binom{R_3}{C},$$

has a zero-one repertory (see Section 2.2). Further, we assume that the confronting system C, which forms part of the model, always "follows the majority" — that is, that the determinator (see Section 1.9) of the throughput

$$\binom{C}{D}$$

can be expressed by the table below (7.2.0).

TABLE 7.2.0

Input			Output
I	II	III	
0	0	0	0
0	0	1	0
0	1	0	0
0	1	1	1
1	0	0	0
1	0	1	1
1	1	0	1
1	1	1	1

It can easily be shown that the system C can be built by coupling zero-one systems of the types which are already familiar to us (Sections 2.2 and 2.3). The details are shown at Fig. 7.2.1.

FIGURE 7.2.1

A parallely confronting system

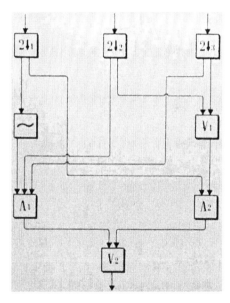

The model consists of eight elementary zero-one systems of five different types (see Table 7.2.2).

The determinators of the five types of systems are described in tabular form in Tables 7.2.3a-7.2.3e.

We now proceed to examine the functioning of direct serial couplings in the model. To begin with

TABLE 7.2.2

(zero-one systems coupled into a parallely confronting system)

No.	Type of systems	List of systems	Number of inputs outputs per system		Number of systems
			inputs	outputs	
1	Duplicating systems	$2\downarrow_1$, $2\downarrow_2$, $2\downarrow_3$	1	2	3
2	Negation system	$\sim$	1	1	1
3	Alternative systems	$\vee_1$, $\vee_2$	2	1	2
4	Conjunction system	$\wedge_1$	3	1	1
5	Conjunction system	$\wedge_2$	2	1	1
			Total of systems		8

the upper "stratum" (Fig. 7.2.1), the system $2\downarrow_1$ is directly coupled serially with the system $\sim$, and the systems $2\downarrow_2$ and $2\downarrow_3$ with the system $\vee_1$.

TABLE 7.2.3a

(replicating system)

Input	Output
0	*0*
1	*1*

TABLE 7.2.3b

(negation system)

Input	Output
0	*1*
1	*0*

TABLE 7.2.3c

(alternative system)

Input I	Input II	Output
0	0	0
0	1	1
1	0	1
1	1	1

TABLE 7.2.3d

(conjunction system)

Input I	Input II	Output
0	0	0
0	1	0
1	0	0
1	1	1

TABLE 7.2.3e

(conjunction system with three inputs)

Input I	Input II	Input III	Output
0	0	0	0
0	0	1	0
0	1	0	0
0	1	1	0
1	0	0	0
1	0	1	0
1	1	0	0
1	1	1	1

This gives us a system with three inputs and two outputs:

(a) the inputs are those of all the three duplicating systems:

$$2{\downarrow}_1, \ 2{\downarrow}_2, \ 2{\downarrow}_3,$$

(b) the outputs are those of the negation system $\sim$ and the alternative system $\vee_1$.

Thus we have to deal here with two determinators which can be easily described in a tabular form on the strength of 7.2.3a, 7.2.3b and 7.2.3c. The results are given in Table 7.2.4.

Now both outputs of the submodel discussed above will be coupled with the systems $\wedge_1$ and $\wedge_2$, in the way shown in Fig. 7.2.1. This gives a wider

TABLE 7.2.4

Input to system			Output from system	
$2\downarrow_1$	$2\downarrow_2$	$2\downarrow_3$	$\sim$	$\vee_1$
0	0	0	1	0
0	0	1	1	1
0	1	0	1	1
0	1	1	1	1
1	0	0	0	0
1	0	1	0	1
1	1	0	0	1
1	1	1	0	1

TABLE 7.2.5

Input to system			Output from system	
$2\downarrow_1$	$2\downarrow_2$	$2\downarrow_3$	$\wedge_1$	$\wedge_2$
0	0	0	0	0
0	0	1	0	0
0	1	0	0	0
0	1	1	1	0
1	0	0	0	0
1	0	1	0	1
1	1	0	0	1
1	1	1	0	1

submodel, with the same three inputs as it had before, and with two outputs — one of the conjunction system $\wedge_1$ and one of the conjunction system $\wedge_2$. The determinators of the outputs of that wider submodel are shown in Table 7.2.5 (calculations can easily be made on the basis of Table 7.2.4, Table 7.2.3e, Table 7.2.3d and Fig. 7.2.1).

The final step consists in coupling that submodel serially with the alternative system $\vee_2$. This gives the complete model, which is a zero-one system with three inputs and one output. The inputs are those only of the three duplicating systems: $2\downarrow_1$,

$2\downarrow_2$, $2\downarrow_3$. The only output is that of the alternative system $\vee_2$. The determinator of that output is a function of three arguments, which can be presented in tabular form on the basis of Table 7.2.5, Table 7.2.3c, and Fig. 7.2.1 (see Table 7.2.6).

TABLE 7.2.6

Input to system			Output from system
$2\downarrow_1$	$2\downarrow_2$	$2\downarrow_3$	$\vee_2$
0	0	0	0
0	0	1	0
0	1	0	0
0	1	1	1
1	0	0	0
1	0	1	1
1	1	0	1
1	1	1	1

If we compare Table 7.2.6 with the table which describes the three-input zero-one system parallely confronting qualitative information (Table 7.2.0), we see that the system shown in Fig. 7.2.1 is precisely the confronting system *C*.

7.3. The binary system

Before modelling computations, we go back to what we were taught at school. We were taught to write natural numbers in the *positional decimal system*, which is based on the following principles:

(1) We write out the first *ten* natural numbers:

$$0, 1, 2, 3, 4, 5, 6, 7, 8, 9.$$

The term "decimal digit" will be applied only to the above numbers.

(2) Every natural number in this system is written in columns (positional system), the columns being denoted (I, II, III, IV, ...) from right to left (which arises from the Semitic order of writing, in the case of figures erroneously adopted in Mediaeval Europe).

(3) Every column is ascribed its unit, which is a consecutive power of ten (the exponent always being the number of the column reduced by unity), namely

$$\text{column} \quad -\text{IV, III, II, I,}$$
$$\text{unit} \quad -X^3, X^2, X^1, X^0,$$

(where $X = \text{ten}$, $X^1 = X$, $X^0 = 1$). Thus the unit of column I is 1, that of column II is X, that of column III is one hundred, that of column IV is one thousand, etc.

(4) Every natural number can be expressed as the sum of products of decimal digit by the units ascribed to columns, viz.:

(a) $C_k \cdot X^{k-1} + C_{k-1} \cdot X^{k-2} + \ldots + C_2 \cdot X^1 \mid C_1 \cdot X^0,$

where $C_1, C_2, \ldots, C_{k-1}, C_k$ are decimal digits.

(5) Instead of writing a given natural number in its long form (a), we write it in a more lucid form:

(b) $\begin{cases} \text{column } k & k-1 & \ldots \text{ II I,} \\ \text{unit} & X^{k-1} & X^{k-2} & \ldots & X^1 \ X^0, \\ \text{digit} & C_k & C_{k-1} & \ldots & C_2 \ C_1; \end{cases}$

or else we write it in an abbreviated form (which is commonly used):

(c) $\qquad\qquad C_k\ C_{k-1}\ \ldots\ C_2\ C_1.$

If the positional decimal system is taken as the basis, a general schema of a positional n-al system (where n is a natural number greater than 1) can very easily be worked out:

(1) We write out n initial natural numbers

$$0, 1, \ldots, n-1,$$

and these numbers, and only these, are called "n-al digits".

(2) Every natural number in this system is written in columns (positional system), the columns being denoted (I, II, III, IV, etc.) from right to left.

(3) Every column is ascribed its unit which is a consecutive power of n (the exponent always being the number of the column less one), namely

$$\text{columns IV III II I}$$
$$\text{units}\quad n^3\ \ n^2\ \ n^1\ n^0$$

(where $n^1 = n$, and $n^0 = 1$).

(4) Every natural number can be written as a sum of products of n-al digits by the units ascribed to columns, namely

(a) $\quad C_k \cdot n^{k-1} + C_{k-1} \cdot n^{k-2} + \ldots + C_2 \cdot n^1 + C_1 \cdot n^0,$

where each of the numbers $C_1, C_2, \ldots, C_{k-1}, C_k$ is an n-al digit.

(5) Instead of writing natural numbers in their long form (a), we can write them in a more lucid form:

(b)
$$
\begin{cases}
\text{column } k & k-1 & \dots & \text{II} & \text{I}, \\
\text{unit} & n^{k-1} & n^{k-2} & \dots & n^1 & n^0, \\
\text{digit} & C_k & C_{k-1} & \dots & C_2 & C_1;
\end{cases}
$$

or else in the abbreviated form:

(c) $$ C_k \; C_{k-1} \; \dots \; C_2 \; C_1. $$

If in the schema described above

$$ n = 2, $$

then we obtain the principles of the positional *binary system*, which has only two digits,

$$ 0, \; 1, $$

already known to us in connection with the zero--one systems (Section 2.2) and the binarization of inputs and outputs (Section 2.5). This fact explains why the binary system has found many applications in cybernetics.

Table 7.3.0 reveals a weak point in the binary system, viz., the length of figures which are recorded in it. In the decimal system, two digits only are needed to write the number ten, but in the binary system four digits must be used for the same purpose. The analogous figures for the number one hundred are three and seven, and four and ten for the number one thousand.

Nevertheless, the strong points of the binary system (e. g., exceptionally simple tables of arithmetical operations — see next section) obviously tip the scales in its favour.

TABLE 7.3.0

Roman system	Positional systems	
	decimal	binary
	0	0
I	1	1
II	2	10
III	3	11
IV	4	100
V	5	101
VI	6	110
VII	7	111
VIII	8	1 000
IX	9	1 001
X	10	1 010
...	...	...
C	100	1 100 100
...	...	...
M	1 000	1 111 101 000

The negative binary system, introduced recently and first described by two young Polish scientists, Z. Pawlak and A. Wakulicz (cf. *Ref.* PW), takes consecutive powers of a negative number, i. e., minus two (−2), as the units ascribed to consecutive columns. Because of its value in the construction of programmed digital computers, it will probably compete with the standard binary system.

7.4. Arithmetical operations

All practical arithmetic (e. g., four elementary operations on natural numbers) lends itself very well to formalization. The language in which the equations forming the tables of addition, multiplication, subtraction and division of natural numbers, as well as the results of such elementary calculations are written down, can be treated as a formal language; such elementary equations can be treated as singled out expressions (Sections 6.9 and 6.10). This fact greatly facilitates the modelling of arithmetic computations and the building of automatic computers (Section 6.11).

It is particularly easy to model arithmetic operations on natural numbers in the binary system (see preceding section). Since any operation to be modelled must be familiar to the modeller, the principles of the arithmetic of natural numbers recorded in the binary system will be outlined here. They will prove to be much simpler than in the case of the decimal system. We begin with binary one-digit numbers:

(1) Addition table for two binary one-digit numbers:

$$0+0 = \ \ 0 = 00,$$
$$0+1 = \ \ 1 = 01,$$
$$1+0 = \ \ 1 = 01,$$
$$1+1 = 10 = 10.$$

(2) Multiplication table for two binary one-digit numbers:

$$0 \cdot 0 = 0,$$
$$0 \cdot 1 = 0,$$
$$1 \cdot 0 = 0,$$
$$1 \cdot 1 = 1.$$

(3) Subtraction table for two binary one-digit numbers:

$$0 - 0 = 0,$$
$$1 - 0 = 1,$$
$$1 - 1 = 0.$$

(4) Division table for two binary one-digit numbers:

$$0 : 1 = 0,$$
$$1 : 1 = 1.$$

To add two binary many-digit numbers we usually need an addition table for *three* binary one-digit numbers:

$$0 + 0 + 0 = 0 = 00,$$
$$0 + 0 + 1 = 1 = 01,$$
$$0 + 1 + 0 = 1 = 01,$$
$$0 + 1 + 1 = 10 = 10,$$

$$1 + 0 + 0 = 1 = 01,$$
$$1 + 0 + 1 = 10 = 10,$$
$$1 + 1 + 0 = 10 = 10,$$
$$1 + 1 + 1 = 11 = 11.$$

To multiply a binary many-digit number by a binary many-digit number we need:

(1) multiplication table of binary one-digit numbers;

(2) a method of doubling (doubling is achieved by writing a zero on the right of the binary number being doubled);

(3) addition table for two binary many-digit numbers.

Here is an example of addition of two binary three-digit numbers:

To add 101+111 we write both factors one under the other:

Column	IV	III	II	I
carried forward	...	...	...	...
1st summand	...	1	0	1
2nd summand	...	1	1	1
sum	...	...	...	...
to be carried forward	...	...	...	...

The figure to be carried forward in column I is, of course, zero, since we have just started to add; consequently, addition in column I is as follows:

$$0 + 1 + 1 = 10 \text{ (see (5))},$$

or

	Column I
carried forward	0
1st summand	1
2nd summand	1
I digit of the sum	0
to be carried forward	1

Since the figure to be carried forward from column I is 1, addition in column II is as follows:

$$1+0+1 = 10 \text{ (see (5))},$$

or

	Column II
carried forward	1
1st summand	0
2nd summand	1
II digit of the sum	0
to be carried forward	1

The figure carried forward from column II being 1, addition in column III is as follows:

$$1+1+1 = 11 \text{ (see (5))},$$

or

	Column III
carried forward	1
1st summand	1
2nd summand	1
III digit of the sum	1
to be carried forward	1

Since the figure carried forward from column III is 1, addition in column IV remains to be performed, i. e.,

$$1+0+0 = 1 \text{ (see (5))},$$

or

	Column IV
carried forward	1
1st summand	0
2nd summand	0
IV digit of the sum	1
to be carried forward	0

The four additions above can be summarized thus:

Column	IV	III	II	I
carried forward	1	1	1	0
1st summand	0	1	0	1
2nd summand	0	1	1	1
sum	1	1	0	0
to be carried forward	0	1	1	1

Thus, $101 + 111 = 1100$. The job is done. The reader is encouraged to try multiplication of many--digit figures.

7.5. A serially adding system

Fig. 7.5.0 shows a serially adding model working in the binary system. The model is a zero-one system (Section 2.2), built only of the elementary zero-one systems familiar to us.

Table 7.5.1 gives a detailed list of the 17 elementary systems involved, which are of seven different kinds. This makes the present model more complicated than any we have dealt with so far. The determinators of the various systems are shown in Tables 7.5.2a to 7.5.2f.

We shall now examine the direct serial couplings uniting these 17 systems into a single model.

First, we must pay attention (Fig. 7.5.0) to the direct couplings of the three quadruplicating systems

$$4\downarrow_1, \quad 4\downarrow_2, \quad 4\downarrow_3$$

FIGURE 7.5.0

A serially adding system

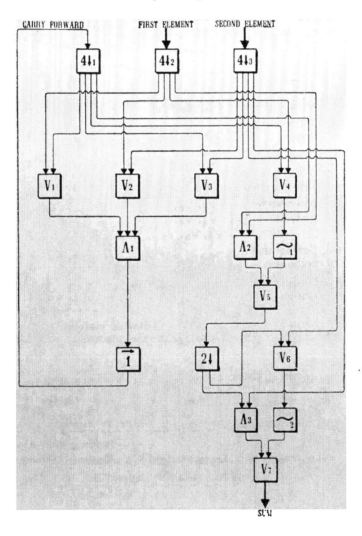

TABLE 7.5.1

(Zero-one system coupled into a serially summing system)

No.	Type of system	List of systems	Number of inputs per system	Number of outputs per system	Number of systems
1	Quadruplic-ating systems	$4\downarrow_1, 4\downarrow_2, 4\downarrow_3$	1	4	3
2	Duplicating system	$2\downarrow$	1	2	1
3	Negation systems	$\sim_1, \sim_2$	1	1	2
4	Delay system	$\vec{1}$	1	1	1
5	Alternative systems	$V_1, V_2, V_3, V_4, V_5, V_6, V_7$	2	1	7
6	Conjunction systems	$\wedge_2, \wedge_3$	2	1	2
7	Conjunction system	$\wedge_1$	3	1	1
				Total of systems	17

TABLE 7.5.2 a	TABLE 7.5.2 b	TABLE 7.5.2 c
(replicating system)	(negation system)	(delay system)

Input	Output
0	*0*
1	*1*

Input	Output
0	*1*
1	*0*

Input at Moment I	Output at Moment II
0	*0*
1	*1*

TABLE 7.5.2 d
(alternative system)

Input I	Input II	Output
0	0	0
0	1	1
1	0	1
1	1	1

TABLE 7.5.2 e
(conjunction system)

Input I	Input II	Output
0	0	0
0	1	0
1	0	0
1	1	1

TABLE 7.5.2 f
(conjunction system with three inputs)

Input I	Input II	Input III	Output
0	0	0	0
0	0	1	0
0	1	0	0
0	1	1	0
1	0	0	0
1	0	1	0
1	1	0	0
1	1	1	1

with the following seven systems:

$$\vee_1, \quad \vee_2, \quad \vee_3, \quad \vee_4, \quad \vee_6,$$

$$\wedge_2, \quad \wedge_3.$$

The determinators of the outputs concerned are shown in Table 7.5.3.

Table 7.5.4 shows the determinators of the outputs of a wider submodel, which, further, includes the systems

$$\wedge_1, \quad \wedge_2, \quad \sim_1.$$

TABLE 7.5.3

Input to system			Output from system							
$4{\downarrow}_1$	$4{\downarrow}_2$	$4{\downarrow}_3$	$\vee_1$	$\vee_2$	$\vee_3$	$\vee_4$	$\binom{4{\downarrow}_1}{\wedge_2}$	$\binom{4{\downarrow}_2}{\wedge_2}$	$\binom{4{\downarrow}_3}{\vee_6}$	$\binom{4{\downarrow}_3}{\wedge_3}$
0	0	0	0	0	0	0	0	0	0	0
0	0	1	0	1	1	0	0	0	1	1
0	1	0	1	1	0	1	0	1	0	0
0	1	1	1	1	1	1	0	1	1	1
1	0	0	1	0	1	1	1	0	0	0
1	0	1	1	1	1	1	1	0	1	1
1	1	0	1	1	1	1	1	1	0	0
1	1	1	1	1	1	1	1	1	1	1

TABLE 7.5.4

Input to system			Output from system				
$4{\downarrow}_1$	$4{\downarrow}_2$	$4{\downarrow}_3$	$\wedge_1$	$\wedge_2$	$\sim_1$	$\binom{4{\downarrow}_3}{\vee_6}$	$\binom{4{\downarrow}_3}{\wedge_3}$
0	0	0	0	0	1	0	0
0	0	1	0	0	1	1	1
0	1	0	0	0	0	0	0
0	1	1	1	0	0	1	1
1	0	0	0	0	0	0	0
1	0	1	1	0	0	1	1
1	1	0	1	1	0	0	0
1	1	1	1	1	0	1	1

We now add to the submodel

(a) the delay system $\overrightarrow{1}$;

(b) the alternative system $\vee_5$.

(Fig. 7.5.0, as before). Our submodel now has three inputs and four outputs, viz.:

(1) inputs to the quadruplicating systems

$$4\downarrow_1, \; 4\downarrow_2, \; 4\downarrow_3,$$

(2) outputs from

(a) the delay system $\begin{pmatrix} \overrightarrow{1} \\ 4\downarrow_1 \end{pmatrix}$;

(b) the alternative system $\begin{pmatrix} \vee_5 \\ 2\downarrow \end{pmatrix}$;

(c) the quadruplicating system $\begin{pmatrix} 4\downarrow_3 \\ \vee_6 \end{pmatrix}$; and

(d) the quadruplicating system $\begin{pmatrix} 4\downarrow_3 \\ \wedge_3 \end{pmatrix}$.

The determinators of these outputs are given in Table 7.5.5.

TABLE 7.5.5

Input to system			Output from system			
$4\downarrow_1$	$4\downarrow_2$	$4\downarrow_3$	$\overrightarrow{1}$	$\vee_5$	$\begin{pmatrix} 4\downarrow_3 \\ \vee_6 \end{pmatrix}$	$\begin{pmatrix} 4\downarrow_3 \\ \wedge_3 \end{pmatrix}$
at Moment I			at Moment II	at Moment I		
0	0	0	0	1	0	0
0	0	1	0	1	1	1
0	1	0	0	0	0	0
0	1	1	1	0	1	1
1	0	0	0	0	0	0
1	0	1	1	0	1	1
1	1	0	1	1	0	0
1	1	1	1	1	1	1

The next step consists in adding two more elementary systems:

(a) the quadruplicating system $4\downarrow_1$ (the delay occurring in the feedback now formed must be taken into consideration),

(b) the duplicating system $2\downarrow$.

Consequently, the submodel now has two inputs, i. e., to the quadruplicating systems

$$4\downarrow_2, \ 4\downarrow_3,$$

four outputs:

$$\begin{pmatrix} 2\downarrow \\ \wedge_3 \end{pmatrix}, \quad \begin{pmatrix} 4\downarrow_3 \\ \wedge_3 \end{pmatrix}, \quad \begin{pmatrix} 2\downarrow \\ \vee_6 \end{pmatrix}, \quad \begin{pmatrix} 4\downarrow_3 \\ \vee_6 \end{pmatrix}, \text{ and}$$

two self-couplings:

(a) $$\begin{pmatrix} 4\downarrow_1 \\ \vee_1 \end{pmatrix}, \quad \begin{pmatrix} \vee_1 \\ \wedge_1 \end{pmatrix}, \quad \begin{pmatrix} \wedge_1 \\ \vec{1} \end{pmatrix}, \quad \begin{pmatrix} \vec{1} \\ 4\downarrow_1 \end{pmatrix},$$

(b) $$\begin{pmatrix} 4\downarrow_1 \\ \vee_3 \end{pmatrix}, \quad \begin{pmatrix} \vee_3 \\ \wedge_1 \end{pmatrix}, \quad \begin{pmatrix} \wedge_1 \\ \vec{1} \end{pmatrix}, \quad \begin{pmatrix} \vec{1} \\ 4\downarrow_1 \end{pmatrix}.$$

The determinators of such outputs are given in Table 7.5.6.

If we now add the next two elementary systems (Fig. 7.5.0):

(a) the conjunction system $\wedge_3$, and

(b) the alternative system $\vee_6$,

then the inputs and the self-couplings of the submodel will undergo no change, but the number of outputs will be reduced to two:

$$\begin{pmatrix} \wedge_3 \\ \vee_7 \end{pmatrix}, \quad \begin{pmatrix} \vee_6 \\ \sim_2 \end{pmatrix}.$$

TABLE 7.5.6

Input to system			Output from system				
$4\downarrow_1$	$4\downarrow_2$	$4\downarrow_3$	$4\downarrow_1$	$\binom{2\downarrow}{\wedge_3}$	$\binom{4\downarrow_3}{\wedge_3}$	$\binom{2\downarrow}{\vee_6}$	$\binom{4\downarrow_3}{\vee_6}$
at Moment I			at Moment II	at Moment I			
0	0	0	0	1	0	1	0
0	0	1	0	1	1	1	1
0	1	0	0	0	0	0	0
0	1	1	1	0	1	0	1
1	0	0	0	0	0	0	0
1	0	1	1	0	1	0	1
1	1	0	1	1	0	1	0
1	1	1	1	1	1	1	1

The determinators involved are shown in Table 7.5.7.

TABLE 7.5.7

Input to system			Output from system		
$4\downarrow_1$	$4\downarrow_2$	$4\downarrow_3$	$4\downarrow_1$	$\wedge_3$	$\vee_6$
at Moment I			at Moment II	at Moment I	
0	0	0	0	0	1
0	0	1	0	1	1
0	1	0	0	0	0
0	1	1	1	0	1
1	0	0	0	0	0
1	0	1	1	0	1
1	1	0	1	0	1
1	1	1	1	1	1

We now take into the submodel the negation system $\sim_2$. This affects one of the outputs of the submodel, which now are

$$\begin{pmatrix} \wedge_3 \\ \vee_7 \end{pmatrix}, \quad \begin{pmatrix} \sim_2 \\ \vee_7 \end{pmatrix}.$$

For the determinators see Table 7.5.8.

TABLE 7.5.8

Input to system			Output from system		
$4\!\downarrow_1$	$4\!\downarrow_2$	$4\!\downarrow_3$	$4\!\downarrow_1$	$\wedge_3$	$\sim_2$
at Moment I			at Moment II	at Moment I	
0	0	0	0	0	0
0	0	1	0	1	0
0	1	0	0	0	1
0	1	1	1	0	0
1	0	0	0	0	1
1	0	1	1	0	0
1	1	0	1	0	0
1	1	1	1	1	0

The final step is to add the alternative system $\vee_7$, which gives us the full model. Table 7.5.9 gives the determinator of the self-coupling and of the output of the model (i. e., the output of the alternative system $\vee_7$, added in the final step of construction). The reader is now advised carefully to compare the addition table of three binary one-digit numbers (Section 7.4) with Table 7.5.9; he will find that they fully coincide with one another.

TABLE 7.5.9

Input to system			Output from system	
$4\downarrow_1$	$4\downarrow_2$	$4\downarrow_3$	$4\downarrow_1$	$\vee_7$
at Moment I			at Moment II	at Moment I
0	0	0	0	0
0	0	1	0	1
0	1	0	0	1
0	1	1	1	0
1	0	0	0	1
1	0	1	1	0
1	1	0	1	0
1	1	1	1	1

This serially summing system (Fig. 7.5.0) is used in the following way (assuming that it has been actually built on relay or electronic principles):

(1) We take into consideration the two many-digit binary numbers which we wish to add (Section 7.3).

(2) At Moment 0 we check whether the distinguishable state of the input to the delay system $\overrightarrow{1}$ is zero. If so, then:

(3) At Moment I we input to the system $4\downarrow_2$ that digit of the 1st summand which is in column I, and at the same time we input to the system $4\downarrow_3$ that digit of the 2nd summand which is in column I. (At the input of $4\downarrow_1$ zero is carried forward.) We immediately obtain (Fig. 7.5.0):

(a) at the output of the model, i. e., at the output of the alternative system $\vee_7$, the digit belonging to column I of the sum; $\quad\rightarrow$

(b) at the input of the delay system 1, the digit to be carried forward to column II.

(4) At Moment II, the digit to be carried forward from column I (delayed by the system $\vec{1}$) appears at the input of the system $4\downarrow_1$; we input to $4\downarrow_2$ that digit of the 1st summand which stands in its column II; we input to $4\downarrow_3$ that digit of the 2nd summand which stands in its column II. We immediately obtain:

(a) at the output of the model, the digit belonging to column II of the sum;

(b) at the input of the delay system $\vec{1}$, the digit to be carried forward to column III.

(5) The procedure described above is, *mutatis mutandis*, repeated until addition is finished.

The reader is advised to go back to the example of addition of binary many-digit numbers (Section 7.4) and go through that example very carefully, yet *not* with the help of Table 7.5.9, but with the help of Fig. 7.5.0 and the determinators of the elementary zero-one systems. This will be a somewhat laborious, but very instructive exercise.

(Cf. *Ref.* B. 1, G. 2, G. 12, GBM, P. 1)

7.6. Formal translation from one language into another

It has been mentioned above that translation from one language into another can *usually* be treated as a formal operation, but this assertion has not been substantiated in any way. The question is important, since if translation is, as a rule, a formal operation then it can comparatively easily be modelled in inanimate technical matter.

Such modelling is certainly possible, for experiments in mechanical translation have already been made. All this, however, appears mysterious even to educated people, who generally do not fully realize the formal character of the work done by such translating machines. Hence, it may be worth while to discuss the matter in more detail.

Let us imagine a text, written in the language A, which is to be translated into the language B. The translation is to be done by a person who *does not understand* a single expression in the languages A and B, but knows the language C. The translator has at his disposal the following aids:

(1) an A-B dictionary,

(2) a grammar of the language A written in the metalanguage C,

(3) a grammar of the language B written in the metalanguage C.

He will proceed as follows:

(1) He will read a sentence in the language A and (*without* understanding it) will analyse it from the syntactical point of view with the help of the grammar of A, written in the metalanguage C which he understands.

(2) On the basis of the result of his syntactical analysis and the data supplied by the $A-B$ dictionary, he will select from B those expressions

which are to comprise the equivalent of the sentence in question formulated in *B*.

(3) With the help of the grammar of *B*, written in *C* (which he understands), and on the strength of the results of his previous analysis, he will build, out of the selected expressions in *B*, a translation of the sentence in question (originally formulated in *A*).

This explanation is certainly too cryptic to catch the reader's imagination. Hence, four highly simplified samples of such procedure will be given below.

7.7. Sample one

We have to do with three languages: *A*, *B* and *C*.

The language *A* consists of the following simple expressions: *Abari, Amadi, adinir, abikik, ak.*

The language *B* consists of the following simple expressions: *Begaso, Bemado, bedinor, bezikor, bel.*

The *A—B* dictionary is as follows (see Table 7.7.0):

TABLE 7.7.0

A	*B*
Abari	*Begaso*
adinir	*bedinor*
ak	*bel*
Amadi	*Bemado*
abikik	*bezikor*

As the language C (strictly: metalanguage), we use the English language plus the names of every expression belonging to A or B. Such names are obtained by writing a given expression in the quotation marks.

The grammar of A, written in C, consists of five qualification rules and three syntactical rules.

Qualification rules:

7.7.1A "*Abari*" is a noun.

7.7.2A "*Amadi*" is a noun.

7.7.3A "*adinir*" is an intransitive verb.

7.7.4A "*abikik*" is a transitive verb.

7.7.5A "*ak*" is a sentential conjunction.

Syntactical rules:

7.7.6A Every simple sentence including an intransitive verb is built after the schema:

$$\begin{bmatrix} \text{noun} \\ \text{subject} \end{bmatrix} \text{ followed by } \begin{bmatrix} \text{intransitive verb} \\ \text{predicate} \end{bmatrix}.$$

7.7.7A Every simple sentence including a transitive verb is built after the schema:

$$\begin{bmatrix} \text{noun} \\ \text{subject} \end{bmatrix} \text{ followed by } \begin{bmatrix} \text{transitive verb} \\ \text{predicate} \end{bmatrix}$$

$$\text{followed by } \begin{bmatrix} \text{noun} \\ \text{complement} \end{bmatrix}$$

7.7.8A Every compound sentence is built after the schema: simple sentence — conjunction — simple sentence.

The grammar of B also consists of five qualification rules and three syntactical rules.

Qualification rules:

7.7.1B *"Begaso"* is a noun.

7.7.2B *"Bemado"* is a noun.

7.7.3B *"bedinor"* is an intransitive verb.

7.7.4B *"bezikor"* is a transitive verb.

7.7.5B *"bel"* is a sentential conjunction.

Syntactical rules:

7.7.6B Every simple sentence including an intransitive verb is built after the schema:

$$\begin{bmatrix} \text{intransitive verb} \\ \text{predicate} \end{bmatrix} \text{ followed by } \begin{bmatrix} \text{noun} \\ \text{subject} \end{bmatrix}$$

7.7.7B Every simple sentence including a transitive verb is built after the schema:

$$\begin{bmatrix} \text{transitive verb} \\ \text{predicate} \end{bmatrix} \text{ followed by } \begin{bmatrix} \text{noun} \\ \text{subject} \end{bmatrix}$$

$$\text{followed by } \begin{bmatrix} \text{noun} \\ \text{complement} \end{bmatrix}$$

7.7.8B Every compound sentence is built after the schema: conjunction — simple sentence — simple sentence.

Thus we have all the data. Our task now is to translate from *A* into *B* the following text: *Amadi adinir. Abari abikik Amadi ak Abari adinir. Amadi abikik Abari.*

We notice that the first sentence is a simple one and includes an intransitive verb, that the third sentence also is simple but includes a transitive verb, and that the second sentence is a compound one and consists of two sentences, the first of which includes a transitive verb, and the second, an intransitive verb.

To begin with the first sentence, the $A-B$ dictionary gives us the equivalents:

Amadi — Bemado
adinir — bedinor.

A brief consultation of the grammar of B (rules 7.7.2B, 7.7.3B, 7.7.6B) results in the translation:

Bedinor Bemado.

In the case of the second sentence, the equivalents are:

Abari — Begaso
abikik — bezikor
Amadi — Bemado
ak   — bel
Abari — Begaso
adinir — bedinor.

It follows from the grammar of B (rules: 7.7.1B, 7.7.4B, 7.7.2B, 7.7.5B, 7.7.3B, 7.7.7B, 7.7.6B, 7.7.8B) that the translation is:

Bel bezikor Begaso Bemado bedinor Begaso.

In the third case the equivalents are:

Amadi — Bemado
azikik — bezikor
Abari — Begaso.

The grammar of B (rules: 7.7.2B, 7.7.4B, 7.7.1B, 7.7.7B) again tell us that the translation is:

Bezikor Bemado Begaso.

Thus, a text has been translated from A into B, although we do not understand a single word either in A or in B.

The reader may object that the example given above merely proves that translation is a formal operation only if the languages in question are radically simplified. In particular, he may draw attention to two simplifications which disagree with his linguistic experience, viz.:

(1) that both languages have no inflection;

(2) that the $A-B$ dictionary gives one-to-one correspondence, which excludes the possibility that equiform words can have different functions in different contexts.

Consequently, other examples, more complicated and more interesting, but at the same time more difficult, will be given.

7.8. Sample two

This time we have to do with three languages: A', B and C. B and C are the same languages as B and C of the preceding section, and the $A'-B$ dictionary is the same as the $A-B$ dictionary given above. Thus, A' consists of the same expressions as A, and its qualification rules are also the same. The difference consists in the syntactical rules which, formulated in the metalanguage C, are:

7.8.0 The word order in any simple sentence consisting of a noun and an intransitive verb is arbitrary.

7.8.1 The word order in any simple sentence built of two nouns and a transitive verb is arbitrary, but:

(a) the subject takes on the ending "*p*";

(b) the complement takes on the ending "*r*".

7.8.2 Every compound sentence is built after the schema: simple sentence — conjunction — simple sentence.

The reader is now required to translate from A' into B the following text:

Adinir Amadi. Amadir Amadip abikik ak Abari adinir. Abikik Amadip Amadir.

7.9. Sample three

Now we have to do with a new combination of languages, namely: A, B' and C. A and C are the same as before (Section 7.7).

B' includes B as its proper part and consists of the following simple expressions:

(1) four nouns:

    Begaso, Bugaro, Bemado, Bumaro,

(2) two intransitive verbs:

    bedinor, begunor,

(3) two transitive verbs:

    bezikor, begukor,

(4) two conjunctions:

bel, beg.

The syntax of B' does not differ from that of B. The $A-B'$ dictionary is as follows (see Table 7.9.0):

TABLE 7.9.0

A	B'
Abari	(1) *Begaso,* (2) *Bugaro*
adinir	(1) *bedinor,* (2) *begunor*
ak	(1) *bel,* (2) *beg*
Amadi	(1) *Bemado,* (2) *Bumaro*
abikik	(1) *bezikor,* (2) *begukor*

This time, the dictionary does not establish any one-to-one correspondence. So how about a formal translation? Let us come back to the text which has already been translated from A into B:

Amadi adinir. Abari abikik Amadi ak Abari adinir. Amadi abikik Abari.

Even now some sort of a partial translation is still possible: we can make an alternative translation of the first sentence:

Predicate: *"bedinor"* or *"begunor"*.
Subject: *"Bemado"* or *"Bumaro"*.

(The reader can try the remaining two sentences.) But what next? Nothing more can be done for the time being. Well, the reader may retort, then there is nothing to sing a song about. But is he quite right? Our reply will consist of two points.

(1) Even this radically simplified example shows that in this case there is still room for practical applications. Suppose there is a paper on physics to be translated from French into English, and the machine which is to translate it is supplied with a dictionary of the type shown in Table 7.9.0. We receive an incomplete translation which includes an excessive amount of English words. For instance, there was *"cloche"* in the French original, and the English equivalents given by the machine are *"bell/rascal"*. Such an incomplete, machine--made translation is then given to an English physicist (who *does not* know French!), and he simply strikes out the superfluous words, makes some stylistic corrections and an English translation is ready (provided that the French original did not savour too much of a literary style and did not include idiomatic expressions).

(2) The example under discussion in this section does not discredit the assertion that translation is essentially a formal operation. All that we need is a better dictionary. So let us now try to prepare it.

7.10. Sample four

Our task is the same as in the preceding section, but we work out our $A - B'$ dictionary in more detail (see Table 7.10.0).

This dictionary makes it possible for the reader to translate the text in question from A into B'.

TABLE 7.10.0

Language A	Language B'	Necessary and sufficient condition of identity of meanings (formulated in the metalanguage C)
Abari	*Begaso*	In a simple sentence, together with "*adinir*"
	Bugaro	In the remaining cases
adinir	*bedinor*	In a simple sentence, together with "*Abari*"
	begunor	In the remaining cases
ak	*bel*	Between simple sentences the first of which includes a transitive verb
	beg	In the remaining cases
Amadi	*Bemado*	In a simple sentence, together with "*Abari*"
	Bumaro	In the remaining cases
abikik	*bezikor*	In a simple sentence which has "*Amadi*" as subject
	begukor	In the remaining cases

This also seems to confirm the assertion concerning the essentially formal character of the operation of translation.

The reader may object that, when natural languages are concerned, the preparation of a dictionary on the principles demonstrated in Table 7.10.0 would be extremely laborious. And there he is undoubtedly right.

(Cf. *Ref.* M. 1 and W. 3)

7.11. Concluding remarks

It would be worth while to describe the construction of zero-one systems able to make machine translations, and to outline the principles of logical machines, such as the Logical Truth Calculator, built in the United States. All this, however, would go far beyond the space permissible for this long chapter on logical models.

(Cf. *Ref*. B. 1, G. 19, S. 3)

8. ECONOMIC MODELS

8.0. Introductory remarks

The applicability of the cybernetic concepts, defined in the first chapter of this book, to the building of economic models will now be briefly demonstrated. The first step will deal with the modelling of production, consumption and trade, the next will be concerned with the information systems necessary to the modelling of planning and reports, and the last will consist in outlining a model of the national economy. Models of the national economy are obtained by coupling, serially and on the feedback principle, of the partial models mentioned above as the first and the second step.

Dr W. Hagemajer has drawn the author's attention to the fact that in building economic models on cybernetic principles (Chapters 1 and 2) one must distinguish two aspects:

(1) the usefulness of the concepts of relatively isolated system and of coupling;

(2) the usefulness of the concepts of information system and of information coupling.

Ad (1). This is but a matter of a new language applied to economic models. That new language may be useful for teaching purposes, but it introduces no novel conceptual elements.

Ad (2). In this case, new conceptual elements appear: we have here to do with economic models with circulation not only of goods and man-power, but also of information. We may build models which show not only transformation of matter (with the help of certain quantities of man-power), but also transformation of information, as well as models which illustrate the reciprocal influence of these two processes.

(Cf. *Ref.* B. 3, G. 1, G. 11, L. 1 and T. 1)

8.1. Production and consumption

Every factory can be represented as a prospective system, its *physical inputs* being the various factors of production (raw materials, semi-finished goods, power supplied in various forms, and human labour), and its basic *physical output* being the goods produced by that factory. An additional output will consist of those quantities of factors of production which remain when a given production cycle has been completed; that output is at the same time an additional input of the system (self-coupling). Besides physical inputs and outputs, such systems usually include information inputs and outputs. Information inputs include: the input through which the given factory receives instructions belonging to the national plan (this is a specific trait of socialist economy), the input through which it receives credits, etc. Information outputs include the output through which the factory sends

13

reports concerning its activity, the output corresponding to the paying back of credits, etc.

The modelling of consumption largely corresponds to the modelling of production. Every consumer group under investigation can be treated as a prospective system. According to the tasks given to the model in question, these groups may vary from a single family to all consumers within a national economy. The basic physical inputs in this case are those through which a given consumption system is supplied with consumer goods. The basic physical output is the man-power — the result of consumption. There are also self-couplings which correspond to durable consumer goods not used up for the moment, and to the man-power produced by the system and earmarked to cater for the needs of the consumers.

8.2. Trade

Any market with regulated prices can be represented as a prospective system. Its principal inputs are: the input corresponding to the amount of money earmarked for purchases (demand), the input corresponding to the goods available (supply), and the input corresponding to the price-list. Its principal outputs are: the output showing the amount of goods purchased (plus the balance of money unspent for lack of supply), and the output showing the amount of money obtained from sales (plus the goods unsold for the lack of demand).

Usually, there are also one or more information outputs conveying information about the transactions made (reports for the planning commission or the price commission in a socialist system).

Fig. 8.2.0 shows the simplest possible model of a market with regulated prices. That model consists of a coupling of four prospective systems:

   (a) the buyer,
   (b) the seller,
   (c) the price commission,
   (d) the transactions office.

The following three feedbacks can immediately be noticed:

$$
(1) \quad
\begin{pmatrix} \text{The buyer} \\ \text{The transactions} \\ \text{office} \end{pmatrix},
\begin{pmatrix} \text{The transactions} \\ \text{office} \\ \text{The buyer} \end{pmatrix};
$$

$$
(2) \quad
\begin{pmatrix} \text{The seller} \\ \text{The transactions} \\ \text{office} \end{pmatrix},
\begin{pmatrix} \text{The transactions} \\ \text{office} \\ \text{The seller} \end{pmatrix};
$$

$$
(3) \quad
\begin{pmatrix} \text{The price} \\ \text{commission} \\ \text{The transactions} \\ \text{office} \end{pmatrix},
\begin{pmatrix} \text{The transactions} \\ \text{office} \\ \text{The price} \\ \text{commission} \end{pmatrix}.
$$

The reader, already accustomed to reading appropriate graphs, will certainly find other, more complicated loops in this almost trivial model.

FIGURE 8.2.0
Couplings of the transaction system

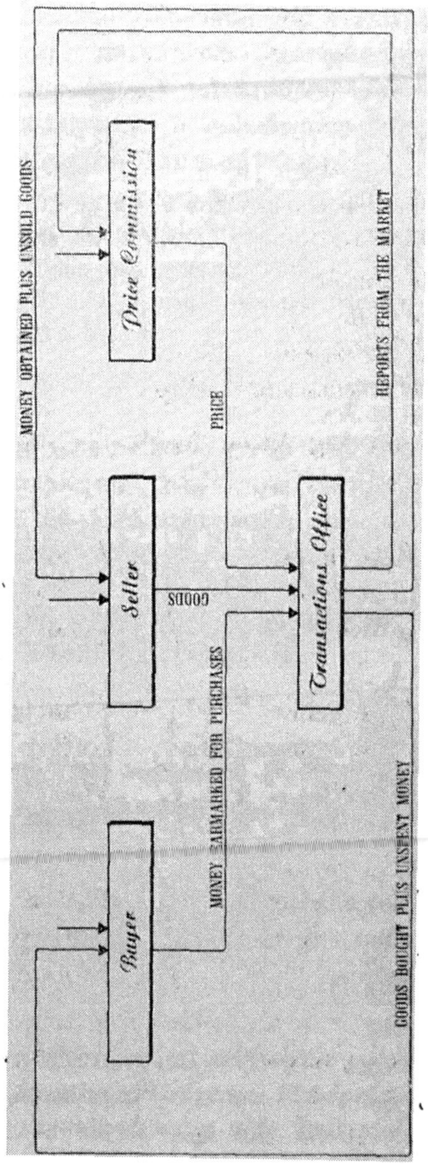

8.3. Planning and reports

Every planning unit, regardless of the scale on which it works, can be represented as a prospective information system: all reports as the inputs of such a system and all instructions at its outputs are treated as information.

In more detailed models, which take into account the fact that the data received in reports are not fully reliable (this refers in particular to the data concerning the implementation of the plan) — i. e., in models which provide for confrontation of contradictory data contained in the reports (see Sections 7.1 and 7.2) — it seems advisable to introduce a separate reporting system, coupled serially with the planning system. Both of them are, of course, information systems (Section 2.0).

Mathematical description (analytical or in the form of matrices) of the determinators of outputs of the reporting or the planning system is possible in the case of models which radically simplify facts, and offers a most promising field for theoretical research. In this connection, a few remarks will be relevant about three simplified types of models of planning systems. They are:

(1) a no-response planning system;

(2) a planning system with non-conditioned responses only;

(3) a planning system with conditioned responses also (Sections 4.1, 4.3).

Ad (1). This means an extremely primitive and "stubborn" system: the determinator of every output of that system is a constant function; the state of inputs has no influence upon the state of any output; throughout the period under investigation, the planning system adheres to one and the same plan and remains deaf to incoming reports. At first glance, this concept seems absurd. And yet it is useful for

(a) teaching purposes,

(b) research purposes.

Ad (a). The rôle of planning in actual socialist economy is so complicated that the only sensible method of teaching consists in consecutive investigation of fictitious planning systems, from absurdly simplified ones to such as resemble the actual state of things.

Ad (b). To show the merits and demerits of a certain plan, it is worth while to study the functioning of a plan which does not change with the lapse of time and remains "deaf" to incoming reports.

Ad (2). This suggests a planning system similar to the automatic pilot. Such a system does not remain deaf to reports, but reacts to them always in accordance with predetermined and unchanging rules.

Ad (3). This means a planning system which is more intelligent than the automatic pilot, since it not only reacts in a standard way to all disequilibrium, but also, after gaining some experience, is able to observe among the data reported certain signals of an approaching disequilibrium and can act preventively.

8.4. Centrally planned economy

By means of serial and feedback couplings of models of production systems, consumption systems, trade systems and suitably chosen information systems (planning and reporting systems), a model of a centrally planned national economy can be obtained.

The building of such a model is a laborious task, and the model itself is a somewhat radical simplification of the actual state of things. Nevertheless, the importance of such models for teaching purposes seems to be beyond question.

We shall confine ourselves here to presenting a model worked out by the Polish Academy of Sciences Econometric Commission (see Fig. 8.4.0).

The principal assumptions referring to the structure of that model are:

(1) Only four kinds of goods are being produced:

(a) perishable producer goods A_0;
(b) durable producer goods A_1;
(c) perishable consumer goods B_0;
(d) durable consumer goods B_1.

(2) Consumption is interpreted as production of man-power C.

(3) The model consists of 17 systems connected by couplings, viz.:

(a) central information systems:

the planning system ABC_*^{I},
the reporting system ABC_+^{I};

FIGURE 8.4.0
Model of a centrally planned national economy

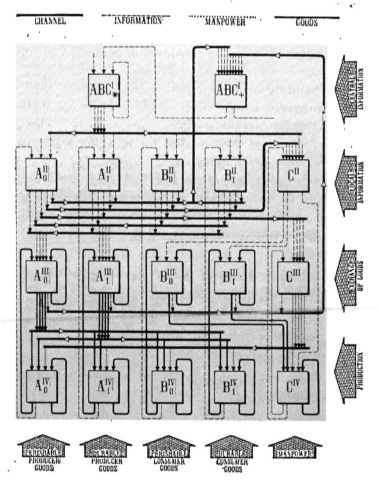

(b) information systems covering individual industries:

A_0^{II} — the budget of production of A_0,
A_1^{II} — the budget of production of A_1,
B_0^{II} — the budget of production of B_0,
B_1^{II} — the budget of production of B_1;

(c) an information system:

C^{II} — the sum of family budgets;

(d) transactions systems:

A_0^{III} — the market of A_0,
A_1^{III} — the market of A_1,
B_0^{III} — the market of B_0,
B_1^{III} — the market of B_1;

(e) a special system:

C^{III} — man-power market;

(f) production systems:

A_0^{IV} — production of A_0,
A_1^{IV} — production of A_1,
B_0^{IV} — production of B_0,
B_1^{IV} — production of B_1;

(g) a special system:

C^{IV} — consumption.

There seems little doubt that the reader who has worked his way through this book will have acquired such experience in discovering serial ·and feedback couplings on the graphs that he can dispense with the author's help in studying the intricate network of couplings in this model of centrally planned national economy.

REFERENCES

A. 1 Ashby W. R. *An Introduction to Cybernetics*. London 1956.
B. 1 Berkeley E. C. *Giant Brains or Machines that Think*. New York 1949.
B. 2 Brillouin L. *Science and Information Theory*. New York 1956.
B. 3 Beach E. F. *Economic Models. An Exposition*. New York 1957.
BGS Bogusławski S., Greniewski H., Szapiro J. *Dialogi o cybernetyce* [Dialogues on Cybernetics]. "Myśl Filozoficzna" no. 4 (14), 1954, pp. 158-212.
C. 1 Choynowski M. *Założenia cybernetyki a zagadnienia biologii* [*The Principles of Cybernetics and Biological Problems*]. "Postępy Wiedzy Medycznej" no. 3, 1957, pp. 239--227.
D. 1 Dembowski J. *Naśladowanie zjawisk życiowych jako metoda biologiczna* [*Imitation of Phenomena of Life as a Biological Method*]. Lwów-Warszawa 1924.
G. 1 Greniewski H. *ABC gospodarki planowej* [*ABC of Planned Economy*]. Warszawa 1947. [Parts I&II].
G. 2 Greniewski H. *Arithmetics of Natural Numbers as a Part of Divalued Propositional Calculus*. "Colloquium Mathematicum" 1951, pp. 291-297.
G. 3 Greniewski H. *Logika matematyczna a sieci elektryczne* [*Mathematical Logic and Electrical Circuits*]. "Problemy" no. 7 (88), 1953, pp. 449-455.
G. 4 Greniewski H. *Elementy logiki formalnej* [*Elements of Formal Logic*]. Warszawa 1955.
G. 5 Greniewski H. *Elementy logiki indukcji* [*Elements of the Logic of Induction*]. Warszawa 1955.

G. 6 Greniewski H. *Milla kanon zmian towarzyszących* [*Mill's Principle of Concomitant Variations*]. "Studia Logica" V, 1957, pp. 109-126.

G. 7 Greniewski H. *Cybernetyka* [*Cybernetics*]. "Encyklopedia Współczesna" no. 3, 1957, pp. 105-107.

G. 8 Greniewski H. *Syntetyczne zwierzęta* [*Synthetic Animals*]. "Encyklopedia Współczesna" no. 10, 1957, pp. 471-474.

G. 9 Greniewski H. *Obrona dysertacji doktorskiej* [*In-Defence of the Doctoral Thesis*]. "Studia Filozoficzne" no. 2 (5), 1958, pp. 237-247.

G. 10 Greniewski H. *Cybernetyka z lotu ptaka* [*Cybernetics — a Bird's-Eye View*]. Warszawa 1959.

G. 11 Greniewski H. *Cybernetics and Economic Models.* "The Reviev of the Polish Academy of Sciences" IV, no. 2 (14), 1959, pp. 57-96.

G. 12 Greniewski M. [junior] *Algebry $(m+n)$-elementowe i ich zastosowania do układów przekaźnikowo-stykowych* [*$(m+n)$-Elemental Algebras and their Application to Relay and Contact Systems*]. "Zastosowania Matematyki" IV, no. 2, 1958, pp. 142-168.

GMB Greniewski H., Bochenek K., Marczyński M. *Application of Bi-elemental Boolean Algebra to Electronic Circuits.* "Studia Logica" II, 1955, pp. 7-76.

K. 1 Kotarbiński T. *Wybór pism. T. I: Myśli o działaniu* [*Selected Works. Vol. 1. Reflections on Action*]. Warszawa 1957.

K. 2 Kuratowski J. *Sur la notion de l'ordre dans la théorie des ensembles.* "Fundamenta Mathematicae" 2, 1921, pp. 116-171.

KM Kuratowski K., Mostowski A. *Teoria mnogości* [*The Set Theory*]. Warszawa-Wrocław 1952.

L. 1 Lange O. *Introduction to Econometrics.* Warszawa-London-New York 1959.

L. 2 Лапунов А. А. *О некоторих общих вопросах кибернетики* [*On Certain General Problems of Cybernetics*]. "Проблемы Кибернетики" 1, 1958, pp. 5-22.

L. 3 Latil P. de *La pensée artificielle.* Paris 1953.

L. 4 Lem S. *Dialogi* [*Dialogues*]. Kraków 1957.

L. 5 Leśniak K. *Filodemosa «Traktat o indukcji»* [*Philo-demos' «Treatise on Induction»*]. "Studia Logica" II, 1957, pp. 77-111.

M. 1 Мельчук И. А. *О машинном переводе с венгерского языка на русский* [*On Mechanical Translation from Hungarian into Russian*]. "Проблемы Кибернетики" 1, 1958, pp. 222-264.

M. 2 Moisil G. C. (Bucureşti) *Zarys algebry automatycznych układów przekaźnikowo-stykowych* [*An Outline of an Algebra of Automatic Relay and Contact Systems*]. "Zastosowania Matematyki" IV, no. 1, 1958, pp. 1-27.

M. 3 Mostowski A. *Logika matematyczna* [*Mathematical Logic*]. Warszawa-Wrocław 1948.

P. 1 Page E. *Digital Computer Switching Circuits.* "Electronics" 7, 1948, pp. 110-118.

PW Pawlak Z., Wakulicz A. *Use of Expansions with a Negative Basis in the Arithmometer of a Digital Computer.* "Bull. Ac. Pol. Sc., III", vol. V, no. 3, 1957, pp. 233-236.

R. 1 Rashevsky N. *Mathematical Biophysics.* Chicago 1948.

R. 2 Russell B. *Mysticism and Logic.* London 1929.

S. 1 Shannon C., Weaver W. *The Mathematical Theory of Communication.* Urbana 1949.

S. 2 Slepian D. *Information Theory.* In: *Operations Research for Management.* Baltimore 1954, pp. 149-167.

S. 3 Sluckin W. *Minds and Machines.* London.

S. 4 Staehler R. E. *An Application of Boolean Algebra to Switching Circuit Design.* "The Bell System Technical Journal" vol. 31, no. 2, 1952, pp. 280-305.

S. 5 Szaniawski K. *O indukcji eliminacyjnej* [*On Eliminative Induction*]. In: *Fragmenty filozoficzne. Seria druga.* Warszawa 1959, pp. 291-306.

S. 6 Szaniawski K. *Prawo, prawidłowość statystyczna, prawdopodobieństwo* [*Scientific Law, Statistical Regularity, Probability*]. "Zeszyty Wydziału Filozoficznego Uniwersytetu Warszawskiego" no. 2, 1957, pp. 73-85.

SKL Соболев С. Л., Китов Е., Ляпунов А. А. *Основные черты кибернетики* [*The Basic Principles of Cybernetics*]. "Вопросы философии" no. 4, 1955, pp. 136-148.

T. 1 Tinbergen J. *Einführung in die Oekonometrie*. Wien-
-Stuttgart 1952.

W. 1 Wiener N. *A Simplification of the Logic of Relations*.
"Proceedings of the Cambridge Philosophical Society"
vol. 17, 1912-1914, pp. 387-390.

W. 2 Wiener N. *Cybernetics, or Control and Communication
in the Animal and the Machine*. New York-Paris 1948.

W. 3 Wojtasiewicz O. *Wstęp do teorii tłumaczenia [An
Introduction to the Theory of Translation]*. Wrocław-War-
szawa 1957.

Date Due

Q310 .G713

Greniewski, Henryk

Cybernetics without
mathematics.

DATE	ISSUED TO
	17389

17389

Q
310
G713

Greniewski, Henryk
    Cybernetics without
mathematics

Trent
University

www.ingramcontent.com/pod-product-compliance
Lightning Source LLC
LaVergne TN
LVHW012203040326
832903LV00003B/86